THE GEOLOGICAL UNCONSCIOUS

The Geological Unconscious

GERMAN LITERATURE AND THE MINERAL IMAGINARY

Jason Groves

FORDHAM UNIVERSITY PRESS   NEW YORK 2020

Copyright © 2020 Fordham University Press

All rights reserved. No part of this publication may be reproduced, stored in a retrieval system, or transmitted in any form or by any means—electronic, mechanical, photocopy, recording, or any other—except for brief quotations in printed reviews, without the prior permission of the publisher.

Fordham University Press has no responsibility for the persistence or accuracy of URLs for external or third-party Internet websites referred to in this publication and does not guarantee that any content on such websites is, or will remain, accurate or appropriate.

Fordham University Press also publishes its books in a variety of electronic formats. Some content that appears in print may not be available in electronic books.

Visit us online at www.fordhampress.com.

Library of Congress Cataloging-in-Publication Data available online at https://catalog.loc.gov.

Printed in the United States of America
22 21 20   5 4 3 2 1
First edition

for Sophie

Contents

Introduction 1

1 Of Other Petrofictions: Reimagining the Mine in German Romanticism 17

2 Goethe's Erratics: Wandering in Deep Time 36

3 Many Stranded Stones: Stifter's Spectral Landscapes 67

4 The Shock of the Earth: Benjamin's Unarticulated Ground 93

Epilogue: Dilapidated 115

ACKNOWLEDGMENTS 139

NOTES 143

BIBLIOGRAPHY 157

INDEX 171

THE GEOLOGICAL UNCONSCIOUS

Introduction

Taking stock of ongoing perturbations to the planet, environmentalist Bill McKibben has argued that the scale of damage renders even a stable reference to the Earth untenable. "We imagine," he writes, "that the disturbances we see around us are the old random and freakish kind. But they're not. It's a different place. A different planet. It needs a new name. Eaarth. Or Monnde, or Tierre, Errde, оккучивать."[1]

With *Eaarth*, McKibben proposes to register these systemic alterations in a defamiliarizing erratum before proceeding to advocate for more sustainable societies better equipped to "make a life on a tough new planet," as the subtitle runs. But perhaps it would be more accurate to recognize not only that the planet is no longer "that stable, secure place" depicted in the iconic 1968 "Earthrise" photograph but also that it was never entirely so.[2] Even when current disturbances are unprecedented and exceptional, the image of the stable, secure planet that is projected onto the past is not entirely justified nor does it always align with the images that circulated during periods of perceived stability. Part of the important work of disturbing the fantasy of the planet as a familiar home must involve unearthing those imaginaries in which other earths emerged.

Even though no national literary tradition can claim to have exclusively anticipated the breadth and magnitude of planetary transformations currently underway, I will focus mostly on German-language writers of imaginative fiction who already in the nineteenth century seriously contended with the challenge of imagining and accounting for a surprisingly volatile planet that bore little resemblance to the images produced by their predecessors. They did so by attending to the instability of the ground, and not only in terms of tectonic activity. The wanderers, wayfarers, and itinerants that populate the literature of this period are often of a mineral nature, and their movements index a kind of planetary turbulence that cannot be measured seismographically. Obtuse, uncongenial, and reticent, in turn ambulatory and sedentary, lively though nonliving, monolithic and clastic, these wanderers have drawn far less critical attention than their fellow bipedal travelers. Such erratics are a primary subject of this study, and the ubiquity of errant earth in modern German literature makes it an apt archive for an exploration of earlier glimmerings of the aberrant planet now identified as Eaarth.

Several historical factors can help to account for the formation of this rich archive. The first modern mining academy was opened in Freiberg in 1756, and several prominent nineteenth-century German-language writers and thinkers (Franz von Baader, Novalis, Theodor Körner, Alexander von Humboldt, Gotthilf Heinrich von Schubert, and Heinrich Steffens) studied at this academy; the lithological theories of its director, Abraham Gottlob Werner, exerted a lasting influence over many writers of this period (most notably, Goethe); and other prominent romantic authors (Clemens Brentano, Joseph von Eichendorff) studied mining at university. The familiarity of these prominent nineteenth-century German-language writers with mining not only meant that they were conversant with the early earth sciences but also that their writings would furnish countless images of mineral matter on the move. The displaced earth that mobilized this mineral imaginary, however, is drawn not only from the manufactured environment of the mine. The free-standing granite blocks in the Northern European flatlands that in the early to mid-nineteenth century earned the epithet of "foundlings" and "erratics" for their stranded and displaced appearance equally piqued the interest of writers and artists contending with the many forms of alienation—social, economic, political—that mark the experience of modernity in German lands. As their visibility increased and as they became less mythical and more epistemological objects, these eccentric foundlings also upset reigning assumptions about the age and behavior of the Earth while becoming some of the earliest sites of land conservation in Switzerland.

My claim here is not that German and German-language writers discovered the earliest signs of what McKibben calls a "new planet" or, more problematically, "the new world"—though as Susanne Zantop has shown, precolonial Germany was shot through with colonial fantasies of discovery and exploration—but rather that the settings of their narratives of wandering proved to be unreliable in ways that later readers recognized as both poetologically and geologically interesting. Although they may not have set out to portray an altered planet, they inadvertently stumbled over those geological and ideological unconformities that also triggered paradigm shifts in scientific discourses. At the same time it bears recognizing that the geocultural context in which these writings take place encompasses those very developments—pioneering mining technologies, advanced geological inquiry and mapping, and scientific exploration in overseas empires marked by dispossession and depopulation, "latent" as well as "manifest" colonialism—that a number of scholars implicate in that destabilization of the Earth system contentiously diagnosed as the "Anthropocene."[3] If the imagination of a volatile planet and its susceptibility to significant climatic shift manifests itself in these texts, an intimation of an imminent destabilization, however, is much harder to discern, let alone its implication in these literary scenes of geological inquiry, resource extraction, and colonial advocacy. Zantop writes of Germany's colonial ambitions that, in the end, reality caught up with the imagination. If the same could be said of Germany's geoengineering ambitions, it is also true that their trajectory from imagination to reality will be far more convoluted. Nevertheless, what follows also includes an attempt to discern a latent Anthropocene in nineteenth-century German-language fiction.

An Inclination Toward Stone

In the fall of 1828, over a period of six weeks, a seventy-five-ton block of granite traveled overland from the Rauen Hills outside of Berlin to the Spree River and from there on a specially designed watercraft to its current location before the Old National Gallery in Berlin. This massive undertaking, whose earliest stages are described by Goethe in an 1828 essay, "Granite Works in Berlin," was perhaps the most spectacular stage in the fabrication of what was the then-largest basin of its kind in the world. Prior to its journey, a massive free-standing granite block known as the Markgrafenstein had been selected from a hilly region near Berlin where such blocks were known to be found, and a segment was blasted out of the rock by hammer, shaped on site, and then prepared for transport. After the six-mile multimodal journey, the basin-shaped block was machine-polished for two years and then, since it could not fit into its intended location in the museum, placed in the courtyard before the museum where it remains today.

Not the stone that the builders took to Berlin but the one that they left behind in the forest became a cornerstone for a new way of thinking and writing about the planet. The story of the basin's travels, as Goethe suspected but never confirmed, extends far beyond its Prussian itinerary and deep into geologic history, in which the Markgrafenstein itself traveled from thousands of miles away, via an ice sheet advancing south from Sweden, to its temporary resting place in the hills outside of Berlin. An encounter with similar blocks may have led Goethe to abandon work on his *Cosmic Novel* but may have also informed his 1829 novel, *Wilhelm Meisters Wanderjahre oder die Entsagenden* (*Wilhelm Meister's Journeyman Years or the Renunciants*). Part of the story that I want to tell in this book is how these adventuresome stones troubled Goethe's writing in scarcely acknowledged ways. More broadly, it is about how such wayward stones perturb narrative in the Age of Goethe and continue to disturb and captivate today. Like Tieck, Stifter, Benjamin, Brecht, and other writers explored in the following pages, throughout his literary career Goethe becomes less concerned with monumental artistic works in stone—and still less with the sublime summits, volcanoes, and earthquakes that pervade the contemporary environmental imagination—and far more with scattered and apparently superfluous stuff, like those Northern German blocks that already by the nineteenth century had been mostly cleared from the landscape and broken down to make aggregate for building projects. These unconsolidated, the aggregate, and the clastic materials become not only the object of a "passionate interest in rocks" (*Neigung zum Gestein*) but also a multifaceted inclination toward stone.[4] Goethe's *Neigung* also disturbs, in addition to the prevailing geo-aesthetics of verticality, the image of the masculinist mountaineer that marks his earlier writings on granite. If not fully adopting the maternal position that informs Adriana Cavarero's attempt to repair the identification of the human figure with uprightness in *Inclinations*, Goethe's own inclinations toward the foundlings point toward a more relational ethical position that for Caverero is marked by care, exposure, and vulnerability.

The Geological Unconscious traces the contours of what could be called, after poet Paul Celan, the "angles of inclination" of a constellation of German-language writers toward stone and the geologic. For Celan, as he remarks in his 1960 Büchner Prize speech, the vitality of poetry can be measured by the degree to which the writer overcomes the artifice of art and writes "according to the angle of declination [*Neigungswinkel*] of his being, the angle of declination of his creatureliness."[5] *Neigungswinkel* refers, among other things, not only to the angular "declination" or deviation between geographic North and magnetic North—as given by a compass reading and as signaled by the title of Celan's address, *The Meridian*—but also to the "dip" or deviation of a geologic stratum from the horizontal plane. In this way the term points to both a geodetic and a geologic aspect of self-understanding and literary production. Creatureliness, Celan seems to suggest, is not only a matter of the human's animality but also its minerality. The unique bent of a life, and accordingly the unique trajectory of a text, is derived from personal as well as impersonal circumstances, deposited by both genealogic and geologic forces. Celan, like the other writers considered here, attempts to register the latter, no matter how faint those traces may be, and no matter how much the forms of inquiry vary. For as much as Celan maintains that the two-fold orientation that characterizes the event of literature takes place in verse form, the inclinations that he describes cannot be circumscribed by verse. The uncongenial stones and inorganic matter considered in the following pages tend to be conveyed in a prosaic language that belies their aberrant trajectories while allowing for their expansive histories.

A turn toward stratigraphy—stratigraphic perspectives, maps, and knowledge—marks early nineteenth-century print culture in Germany. As Andrew Piper has shown, during this time "the notion of a 'deep earth'" was widely disseminated and thereby "established the pictorial scene against which the deep self of romanticism" defined itself; Goethe's collaboration in 1821 on the first geological map of Germany, a work that shares with the 1821 edition of his novel *Journeyman Years* an interest in precipitous topographies, contributed to the widespread "geologization of vision and subjectivity" that contemporary culture inherits.[6] These cartographic visualizations and literary descriptions may have been derived from the findings of the emerging earth sciences, but, as Georg Braungart argues, the contribution of imaginative literature to scientific discourse is more than merely illustrative. Tracing out the methodological interrelation between the literary imagination and geoscientific reconstruction of the planet's past and drawing on Joseph Vogl's notion of a "poetology of knowledge" in which the literary text is a "generator of knowledge" rather than "a reiteration of knowledge produced somewhere else,"[7] Braungart also attests to a complementarity between literature and geology. This complementarity is elaborated by Noah Heringman as an aesthetic geology that occupies a space between literature and geology and concisely summed up by Adelene Buckland in her recent assessment of a literary aspect of nineteenth-century British geology: "Doing geology meant writing it too."[8]

The Geological Unconscious diverges from these studies on at least two fronts. Its lines of inquiry deviate from the focus on the invention of modern geology that characterizes many existing studies. Instead it charts a loosely assembled but at times surprisingly intimate relationship between what pertains to literature and what pertains to stone, namely the literary and the geologic, as articulated by Elizabeth Ellsworth and Jamie Kruse. Borrowing from and minimally adapting Bruno Latour's redirection of sociology into "the social" in *Assembling the Social*, Ellsworth and Kruse seek to redirect geology into the realm of the geologic. As they write: "Even though most geologists would prefer to call 'the geologic' a homogeneous thing, it's perfectly acceptable to designate by the same word a trail of *associations* between heterogeneous elements. . . . In this meaning of the adjective, geologic does not designate a thing among other things, like a black sheep among white sheep, but a *type of connection* between things that are not themselves geologic."[9]

First, this passage provides an opportunity to clarify and also further muddle a terminological distinction between *geologic* and *geological*. As I am using the terms, *geologic* refers to the physical earth and material processes (e.g., geologic formation, geologic time), whereas *geological* refers to the study of the physical earth (e.g., a geological survey). Unlike popular usage, I will attempt to maintain this distinction wherever possible; however, even here this distinction is complicated by the feedback loops between geological study (and industries) and current geologic changes to the planet. This blurring of the geologic and the geological is furthered by theorists such as Latour, Ellsworth, and Kruse, who seek to show how the geologic (and the geological) is materially informed by social and aesthetic formations. If the above definition of the geologic leans heavily on a social context of geology that becomes most pronounced in the Great Acceleration after World War II, the identification of the geologic in connections between things that are not typically thought of as relating to the study of geology—such as commentaries, essays, poems, novellas, and novels, rather than the maps, surveys, and studies generated by practitioners in the field of geology—is useful to account for the literary inclinations toward the lithic as they take shape in the nineteenth century. The same could be said of the inclination of the lithic toward the literary (e.g., the trope of the "rock record" and the "book of nature") in the aesthetic geology of this period.

Second, rather than the deep earth of stratigraphy, the protagonists considered here define themselves in relation to the turbulent surface of the Earth. What they have in common is an inclination toward the erratic: not only what is stratigraphically deviant but rather what deviates from stratigraphic thinking. Studies of literary landscapes, however, have typically overlooked erratics, not to mention the epistemological blocks and crises they induced. Although Marjorie Nicolson's pioneering 1955 monograph *Mountain Gloom, Mountain Glory* offered a much needed historical account of the shifting reception of geologic formations in British and European art and aesthetics from the mid-seventeenth to the late eighteenth-century, much of the scholarship that has appeared in its wake continued to focus on the mountainous and the monumental, and at the expense of those unmonumental objects and formations that were not only the subject of intense scientific scrutiny but also the instigators and receptors of shifting aesthetic regimes. The ostensibly displaced boulders that became known as erratics—as well as the theories of ice ages, deep time, and climatic change that they helped to instigate—would go on to displace existing theories of the Earth arguably to a greater extent than the most spectacular tectonic event in modern European history, the Lisbon earthquake of 1755. Curiously, the geologic erratics and their outcroppings in imaginative fiction have rarely found their way into literary and cultural scholarship. These studies resemble something of a *terrain erratique*, consisting of miscellaneous articles and chapters that range from literary studies to geography and landscape design. The inevitably erratic lines of inquiry explored in *The Geological Unconscious*, however, neither attempt to unify these disparate forms of inquiry nor to offer a comprehensive, systematic theory of the erratic. Instead, the "erratic" develops beyond the constraints of a naturalistic object in order to support inquiry into a range of aberrant movements of bodies, temporalities, and inscriptions at a nexus of the literary and the lithic. Nevertheless, many of my readings telescope through geologic (and geological) erratics. The motifs of mobility, wandering, and errancy that figure prominently in the romantic, realist, and modernist texts under consideration have as their subtext and can be read as allegories of a lithosphere increasingly agitated not only by the expanding extraction and transportation of mineral resources but also by a rapidly expanding insight into the historically volatile climatic conditions of the planet. This insight is drawn in no small part from observation of the peculiar lithic objects that come to be defined as erratics.

In a word, these stories betray a *geological unconscious*—to borrow from Fredric Jameson's "political unconscious" as well as Shoshana Felman's "juridical unconscious"—as a horizon of reading and interpretation. In opening out onto such a horizon these readings offer an opportunity to take up and extend Cathy Caruth's invitation, posed at the close of *Literature in the Ashes of History*, to consider how stories of trauma "bear witness not simply to the past but also to the pasts we have not known."[10] There is no telling how far back into the past or how belated their witnessing might be. These are histories that are inscribed in mineral components, like ash, which literature takes up and transforms. What it takes up and takes note of does not always become conscious knowledge, but rather an "unthought knowledge," as Gabriele Schwab writes, "[of] experiences that have been lived but never known."[11] Those stories that concern our experience of and historical relation to an altering planet offer some of the more extreme cases of such knowledge. Yet they might also betray a more active unknowing. As Christophe Bonneuil and Jean-Baptiste Fressoz observe, "The history of the Anthropocene is not one of a frenetic modernism that transforms the world while ignorant of nature, but rather of the scientific and political production of a modernizing unconscious."[12] The workings of this modernism are frequently registered in literature, perhaps most strikingly in a period where the Anthropocene, though underway, remains unnamed and unthought.

The Humiliation of Literature

Opening the question of the literary and the geologic in a German-language context thus involves reopening a wound in which psychoanalytic thought is also implicated. Between the second and third of the three great *Kränkungen* (translated as both "wounds" and "blows"[13]) to anthropocentrism that Freud announces in a 1917 essay—the cosmological wound wrought by Copernican heliocentrism and the resulting marginalization of the Earth in the solar system; the biological wound wrought by Darwinism and the resulting evolutionary marginalization of the human species; and finally the psychological wound wrought by the topographic model of Freudian psychoanalysis and the resulting marginalization of consciousness in psychic life—paleontologist Stephen Jay Gould interpolates another: a geological wound resulting from the late eighteenth- and early nineteenth-century discovery of a "deep" planetary history and the resulting temporal marginalization of the species on a planet several billion years old.[14] To defend Freud's oversight by pointing out his proclivity for triads would be beside the point, since his work in fact does not neglect the implications of geology and geological discourse for consciousness and psychoanalysis, as seen in work ranging from stratigraphic modeling of the psyche in *The Interpretation of Dreams* (1900) to tracing the ongoing effects of volcanic trauma in "Delusion and Dream in Jensen's *Gradiva*" (1907) to exploring the inclination toward the inorganic known as the death drive in *Beyond the Pleasure Principle* (1920). He implicitly acknowledges this vulnerability, this geotrauma, even when he fails to name it explicitly in an essay in 1917.[15]

Or rather, this vulnerability only emerges in another scene and in another language. The account of the *Kränkung* given by Timothy Morton in *Hyperobjects* furthers Freud's decentering of human narcissism by translating it as "humiliation" and in the literal sense of the human becoming humus: "being brought low, being brought down to earth."[15] If casting Copernican, Darwinian, and Freudian revolutions as all-too-human "wounds" subtly maintained the anthropocentrism and subjectivization that was supposed to be critically injured, reading them as a literal "humiliation" instead opens out into a more radical marginalization of organic life on the planet and a less anthropocentric concept of the human, one in which a deep affinity between the human and what has been taken to be most foreign to it, the mineral realm that Kant termed the *Menschenfremdeste (what is most foreign to the human)*, becomes apparent.[16]

What Morton's gloss on Freud misses, though, is that this humiliating "quake in being" that Freud perhaps inadvertently announces can be read in part as an aftershock issuing from the German romantic tradition from which he extensively drew and whose imaginative fiction, as Heather Sullivan observes, "undermines the certainty of distinguishing between the organic and the inorganic, and . . . renders the human body dangerously close to the minerals."[17] Even the prominent anthropologist Elizabeth Povinelli acknowledges that, with respect to the increasing difficulty for biology and philosophy to maintain a reliable distinction between life and nonlife, "these academic disciplines are only catching up to a conversation begun in literature."[18] Against the backdrop of this uncertainty another register of Goethe's *Neigung zum Gestein* becomes audible in translation: this "passionate interest in rocks" describes a state of being between the human and the mineral that cannot be entirely localized to the human and that entails a self-alienation of the human (where it is defined as an organic body in distinction to an inorganic one). The ecocritical significance of the literature that elaborates this interest should become clear over the course of this study, but for the moment John Crowe Ransom's gloss on this term can stand as a placeholder: "Inter-esse means to be environed, and interest means sensitiveness to environment."[19] The extent to which this sensitiveness is perceived as a humiliating vulnerability demonstrates the sheer fragility of the image of the human in this literature.

Although the intensity of this unsettling interest and ontological humiliation is arguably most pronounced in the first half of the nineteenth century, with respect to German-language literature, it cannot be contained to any one period, genre, or form. It returns, in other settings and other forms, but perhaps nowhere more insistently than the literary arts, as attested to by a 1958 speech delivered by Romanian-Jewish poet Paul Celan on the occasion of receiving the Bremen literature prize, in which Celan—who had been sent to a forced labor camp from 1942 to 1943, whose parents both died during their internment, and whose subsequent poems, in the words of Theodor Adorno, "imitate a language beneath the helpless language of human beings, indeed beneath all organic language"—testifies to the humiliation of language and literature.[20] In its most memorable passage the speech concerns the fate of language in a time of genocide: "It, language, remained unlost, indeed, in spite of everything. But it had to pass through its own answerlessness, pass through frightful muting, pass through the thousand darknesses of deathbringing speech. It passed through and gave no words for that which happened; yet it passed through this happening. Passed through and could come to light [*zutage treten*] again, 'enriched' ['*angereichert*'] by all that."[21] The passage of language through a time of darkness also entails the traumatic repetition of that darkness in and as language. Its return to light reveals the extent of its petrification: this reemergence of language is rendered as *zutage treten*, which is borrowed from mining vernacular and refers to the "outcropping" of a rock formation such as an ore. Just as the form in which language makes its own resilience manifest, the outcrop, avails itself of a geological figure, so too does the content: "enrichment" broadly refers to the formation of mineral deposits as well as an increase in the proportion of a particular isotope in an element (most conspicuously that of the isotope U-235 in uranium) so as to make that element more powerful or explosive. The enrichment of language entails both its petrification and its increased volatility. This form of enrichment is not always explicit and propositional but instead operates through a volatility of reference, as in the sudden audibility of *Reich* in the German *angereichert* ("enriched"), which alludes to the Third Reich and in this way to the nuclear ambitions and toxic enrichment of the language by that regime.

The two-fold humiliation and enrichment of language and literature to which Celan alludes unexpectedly resonates in the rhetorical thrust of *Eaarth*. McKibben proposes to register the withdrawal of the familiarity of the Earth as a home, habitat, and stable point of reference for environmental thought in a time of global climate change in an elemental *a*ccretion. Like the Anthropocene epithet, "Eaarth" marks these profound and far-reaching geophysical transformations through the transformation of a primary geophysical signifier ("Earth"). If to speak of the Earth today is to speak of this ruptured reference, what does this mean for those stories in which writers try to make sense of their place on the planet? What Jussi Parikka observes in *Geology of Media*— that the scale of geologic durations exposed in the nineteenth century "demands an understanding of a story that is radically different from the usual meaning of storytelling with which we usually engage in the humanities"— seems to pertain all the more today; moreover, Parikka writes, "this story is likely to contain fewer words and more a-signifying semiotic matter."[22] In a provisional response to Parikka, the foregoing and following remarks suggest that already in the nineteenth century writers were seriously contending with the implications of geologic time and humiliation for narrative. Yet this observation does not sufficiently answer the inevitable question: How could the geochronological unit of the Anthropocene (or for that matter the Holocene), already spanning decades if not centuries, provide a perspective in which "literature" would still be legible or legitimate as an expression of the (in)human condition?

According to an array of critical voices, it might not. Tom Cohen argues that the time of the Anthropocene, which for the purposes of this book would encompass the nineteenth century, is not "of literature" but rather of cinematic time and an "inhuman, machinic, and interrupted temporality."[23] Relatedly, Ursula Heise suggests that a "database aesthetic" affords more possibilities to register and represent the current mass extinction of biodiversity than the traditional narratives of nature writing.[24] In an epoch where "the climatic in general and what issues from it" has become, to quote Cohen, "radically counter-linear, a mutating hive of feedback loops of counter-referential force-lines, tensions, transferences," literature appears as anachronism, a medium that is out of time, inapt to convey the spatial and temporal scales in which human activity, to put it imprecisely, now registers itself.[25]

Without dismissing these arguments, I propose a different tack. Instead of determining exactly where or when literature became outstripped by the temporalities of the cinema or the aesthetics of the database, it can be pointed out that the concept of literature—like the concept of nature, the Earth, the human, or the humanities—does not arrive completely intact in the twenty-first century. At least three isotopes of literature have been proposed in recent decades: *loiterature* (Ross Chambers), *lituraterre* (Jacques Lacan), and what Thomas Schestag, glossing a line by Walter Benjamin, refers to as *lithorature*.[26] In ways that remain to be worked out in detail, these isotopes of literature point to a growing entanglement of the lithic and the literary, of litter and letter, in a time where litter is literally undergoing lithification. This entanglement is acutely framed by the Anthropocene and is registered in the progressing withdrawal of an intact reference to the Earth, to a stabile ground, to a tenable concept of nature, to language, to literature, to the human. Yet in what follows I attempt to consider the ongoing "enrichment" of the Eaarth together with the enrichment of literature, the multiple humiliations of the human and the humiliations of the Humanities, as they were emerging well before the Anthropocene became an object of scientific discourse. Moreover, rather than making an ontological turn, toward being brought low, this book makes a geophilological turn toward reading brought low, from the mountain to the moraine, from life to nonlife, to the point where an epigrammatic inscription and a glacial striation might become indistinguishable. Methodologically, what follows is a close reading at earth magnitude: a mode of reading for the planet that draws on a "geomethodology" largely in accordance with Christian Moraru's "critical microscopy" of "little places" and other "geo-cultural minutia" as they promise to telescope the planetary; it is also a mode of reading that, unlike Moraru's rejection of the ecocritical in favor of a cultural-critical "intra-anthropological relatedness," sees even the intra-anthropological as being traversed by the ecological and the inhuman.[27]

Structure of the Book

The Geological Unconscious is driven by inquiry into the literary forms of geologic intimacy that become prominent starting in and around German romanticism. Each chapter explores how an attentiveness to stone, and often to a particular stone, informs a work of literature: how German romanticism is coterminous with a writer's descent into a mine and then how literary images of the mine take on a life of their own; how an encounter with the displaced boulders known as foundlings might have led Goethe to abandon his first novel and then to the formal innovations of his last one; how stones serve as witnesses to historical trauma in Adalbert Stifter's realist novellas and how attempts to downplay those histories only make stone even more menacing, and also how the susceptibility of those stones and histories to erosion informs his stories' susceptibility to revision; how an irregular ground underfoot resonates throughout Walter Benjamin's prose, in response to which Benjamin articulates a form of shock (*Erschütterung*) that might contend with the geologic temporality in which the footprint of the species registers itself today; how Benjamin's commentary on the lapidary style of Brecht's *War Primer* poems, and in particular on a line written in chalk, anticipates the possibilities and challenges of commentary in the generalized state of war that is known as the Anthropocene. This book also explores the humiliation of literature: the literature of nonprecious stones, of foundlings in the flatlands, effaced inscriptions in eroded sandstone, uneven paving stones that expose our awkwardness, anti-war slogans written in a piece of chalk that itself bears traces of ancient extinctions, earth with a lowercase *e*, eaarth as erratum.

The word *mineral* ("something mined") condenses the otherwise spatially and temporally mediated relation between stones and their excavation. German romanticism reconstitutes the myriad relations that have been collapsed in mineral objects and in the process produces such an abundance of images that a mineral imaginary comes into formation. Chapter 1 surveys Ludwig Tieck's *Rune Mountain*, his 1801 tale of an eccentric miner's exposure to a mineral aberration, together with Tieck's well-documented visit with Wackenroder in 1793 to iron ore mines in Upper Franconia that is widely regarded as the basis for *Rune Mountain* as well as the initial document of German romanticism. My impulse to reexamine these crucial texts of German romanticism is two-fold: first, the narrative of descent in the letter presents a literal and figurative act of humiliation coterminous with the emergence of a new literary movement. Second, the prevailing consideration of the mine in German romanticism strictly as a mine of the soul rather than a technological site seems inadequate today. In so far as the lithosphere, and most emphatically the mine, can be regarded as a crucible for emergent images and literary movements alike, this perspective also facilitates the expansion of a geology of media beyond technical media culture to encompass literary media as well.

The human-mineral encounters staged in Tieck's story and those that follow also invite a redefinition of the recently proposed genre of petrofiction to take into account the full range of petric encounters in literature and thereby to elaborate relations to the geologic that are not primarily mediated by fossil fuels. The subsequent chapters also expand the mineral imaginary beyond the mine and the mineral per se, defined as it is by a distinction to rocks, which are composed of aggregated minerals. Although granite (an aggregate rather than a mineral) plays a privileged role in mediating culture for Goethe, the later turn to granite foundlings has received little consideration by his readers. Chapter 2 explores how Goethe's thought takes shape less around the fixed stratifications of granite mountains and more around the errant mobility of granite within a climatologically volatile planet. These blocks are the focus of numerous writings from 1816 to 1831, as well as an entire chapter in the 1829 edition of *Journeyman Years*. By the time of his 1828 essay on an errant boulder outside of Berlin, granite presents a considerable geological and therefore poetological problem, and it invites a rereading of the protagonist's errancy within the context of the shifting surfaces of a wandering planet.

Chapter 3 finds a common ground with the previous one in the role of granite in exploring new forms of perception, namely the perception of an unreliable earth. My reading of Adalbert Stifter's *Granite*, the introductory story from the collection *Many-Colored Stones* (1853), turns from the prominent stone announced in the title and opening lines, but instead focuses on the seemingly superfluous stones scattered throughout the story in an attempt to recuperate the geologic import of the writing that Stifter's friend and noted glaciologist Friedrich Simony once described as "petty detail painting of unimportant things."[28] Where other prominent readers have only observed a mortifying reification in the excessive description of Stifter's literary landscapes, I offer a reading that registers an unsettling mobility and shocking vibrancy both in the debris that marks his geologically active landscapes and in the so-called "discarded pages" that result from his notorious revisionary practices. The image of the Earth that emerges in my reading of *Granite* alongside its earlier iteration *The Pitch Burners* (1848)—the Earth as subject to ongoing alteration—informs and is informed by its epistemological status in geological inquiry around the time of the composition of *Granite*.

Chapter 4 turns to Walter Benjamin (one of Stifter's most outspoken critics) to make sense of this withdrawal of the lithosphere as a reliable setting, unobtrusive backdrop, and stable point of reference for literature; in particular, the chapter explores how this withdrawal of the lithosphere corresponds to the withdrawal of "shock" as a tenable aesthetic response to planetary transformations. In this regard, Benjamin's reading of Stifter (discussed in chapter 3) as someone absolutely lacking the ability to present shock is turned on its head. Stifter's inability or unwillingness to express his shock at a recent revolution, both political and geological, is reclaimed as a virtue; in the process an alternative form of shock elaborated by Benjamin becomes much more interesting. For Benjamin, who is particularly attuned to the "unregulated earth" during his retreat on (and to) the island of Ibiza in 1932, the physical ground becomes the locus of a shocking experience. His walks within and without the city register a geomorphic mode of shock—the tremor known in German as the *Erschütterung*—that promises to better address the representational and conceptual challenges posed by the incremental, often anonymous, and imperceptible geologic changes registered in the previous chapters.

The epilogue turns from aesthetic strategies for making sense of the Anthropocene to reflect on the transformation—in a word the *dilapidation*—of literary-critical practices in a planetary context marked by massive geologic change. To think through the derangements of scale in a time of natural-historical emergency, I turn to Walter Benjamin's 1938 commentary on an epigrammatic poem in "the lapidary style," written by Bertolt Brecht during the terminal years of the Holocene if not also the end of the world, or in Benjamin's term, *Weltuntergang*. The focus on the lapidary in a work that addresses the immense historical and political shifts suggests that the "enlarged temporal and geographical scales" that critics argue are necessary for the "ecological project of thinking beyond anthropocentricity" could still be realizable in small forms and through close readings of the diminutive.[29] In this way Brecht's poem and Benjamin's commentary reinforces the incommensurate significance of the erratic blocks of the previous chapters. For Benjamin the shifting dimensions of the lapidary index planetary change. As he turns toward stone and the style of writing developed for it, he finds only the transient and ephemeral. In this way his remarks unsettle the prevailing geopoetic dichotomy of the lapidary and the seismic, the enduring and the unsettling, respectively. Benjamin's reading of Brecht thereby emerges as a commentary for the *destratifying* moment of the Anthropocene and the becoming-erratic of the lapidary.

For a New Modality of Theory

To speak of the imaginary is to speak doubly, no less in this case than others.[30] One version of a mineral imaginary would correspond to a prevailing definition of a cultural imaginary that would encompass "those vast networks of interlinking, discursive themes, images, motifs, and narrative forms that are publicly available within a culture at any one time, and articulate its psychic and social dimensions."[31] The psychic, social, and cultural dimensions of the mine and mining in German romantic literature certainly constitute a mineral imaginary in this sense. Moreover, the mineral imaginary can help cultivate an alternative to the prevailing geological imaginary that is grounded in extractive practices. *The Geological Unconscious* does not explicitly set out to imagine alternatives to fossil capitalism. Yet in elaborating a range of lithic encounters and in registering libidinal investments in the lithosphere outside of the carboniferous, it points toward alternative relations with and less destructive mobilizations of the geologic.

Literature broadly situated in and around the lithosphere also testifies to the menacing capacity of minerals to form images. This other version of the mineral imaginary is *of the mine* both in terms of the site of its emergence and the material of its composition. This version is still congruous with a romantic view of the imagination as a continuation of natural forces, but it also anticipates a radicalization of this view that comes to the fore in surrealism, as voiced by Roger Caillois's claim that "the imagination is nothing more than an extension of matter."[32] Stones that stare, stones that speak, and minerals that reminisce populate German romantic, realist, and modernist literature. Even though a Lacanian reading can account for these strange stones anthropocentrically with recourse to the externality of the gaze, the inhuman images glimpsed in stories of mineral adventures might also offer a model, no matter how rudimentary, for the "new modality of theory" that critics have sought in the term *Anthropocene* and its ability to generate "viewpoints, framings or intuitions of an inhuman look."[33]

Just as the human is glimpsed through mineral and metallic eyes in Tieck's *Rune Mountain*, in the Anthropocenic perspective the putative *anthropos* who (or rather which) will be theoretically legible in the rock record long after the extinction of every human is also the object of a mineral imaginary in this double sense. That an image of the human, no matter how monstrous, could develop through geologic processes is intuited in those literary depictions of mining also developing out of the very processes of extraction that, in many accounts, triggered planetary change. These stories of human-mineral encounters offer a perspective in which the image of the human cannot be distinguished from the image of the mineral and thus in which the minerality of the human becomes apparent. Practices of reading these images, like the commentaries on the following pages, facilitate those awkward perspectives, as Hölderlin once wrote, in which an unexpected truth emerges, and what truth could be more unexpected, what perspective more awkward, than one in which the mineral envisions while also being envisioned? If the place of literature is to imagine the unimaginable then the mineral imaginary is one which we are obliged to think today.

1
Of Other Petrofictions
Reimagining the Mine in German Romanticism

Literature emerges not only in response to philosophical crises, as Jean-Luc Nancy argues in *The Literary Absolute*, but also to planetary crises. The genre of petrofiction, which was first elaborated in 1992 by Amitav Ghosh to refer to Abdelrahman Munif's *Cities of Salt* and other fiction explicitly confronting the oil encounter between the Americas and the Middle East, is one such example.[1] The extraction, processing, and consumption of petroleum and the tangible effects of those processes on sites of production and transmission and ultimately even on the Earth system has helped usher in the recent recasting of literary and cultural production within the periodizing framework of a petromodernity. However, as petroculture comes to pervade environments and the environmental imagination alike—Heather Sullivan writes that "we might today describe almost all literature from industrialized countries as well as countries with extractive economies as 'petro-texts'"—it becomes easier to lose sight of its cultural and historical specificity.[2] In restricting the range of this term to the encounter with oil, theorists of petrofiction constrain both the meaning of the Greek root *petros* (a free-standing stone) and the broader geological imagination. If one takes into account the wide range of petric encounters in modern fiction, those with both free-standing *petros* and massive *petras* (rock formations such as cliffs, deposits, and bedrock), multiple petrofictions begin to emerge. Petroleum may pervade many contemporary cultures across the planet, yet the significant cultural impacts of other deposits, molten or crystallized, solidifying or liquefying, exposed or subsurface, carboniferous or not, have also been the subject of fictionalized encounters. As the damaging effects of centuries of intensive hydrocarbon mining escalate and as societies become ever more oil-bound, it also becomes increasingly urgent to recognize and elaborate relations to the geologic mediated by objects other than fossil fuels.

Redefining the Petro-text

Among the multiple petromodernities in play—with *petro-* no longer understood exclusively with reference to petroleum—this chapter examines a rich seam in what could be called *German petromanticism*, a lithic-literary subperiod that emerged at the saddle of the eighteenth and nineteenth centuries. The period is typified by genres centered on human-mineral romances and occurring in a larger cultural-historical context of intensive interest in forms of mining, particularly of precious stones, metals, and ores—but not coal and other fossil fuels. During the first decade of the nineteenth-century, production of coal in the German lands was, in Theodor Ziolkowski's estimation, "almost too trivial to be recorded" at 0.3 million tons. Compared to England's eleven million tons, and the relative inaccessibility of Germany's fossil fuel deposits throughout much of the nineteenth century, means that the social context and material content of this literature differs from that of their British contemporaries and of contemporary fiction.[3] That there is no endemic oil culture to speak of during this period, and that fossil fuels played a relatively little role in cultural production, does not mean there is no petrofiction to speak of. The mine has been identified as a primary institution of German romanticism, since from the medieval ages through the mid-nineteenth-century German mines served as the primary European source of precious metals for Europe, and by the late eighteenth-century there is a massive influx of fiction dealing with those mines. Just as Jussi Parikka in *Anthrobscene* (2014) and *A Geology of Media* (2015) proposes to consider "the depths of mines as essential places for the emergence of technical media culture," so too do those depths act as an essential place for the emergence of romantic literary culture.[4] The overlap between the cultural and the geological in German romanticism is so pronounced that a letter detailing an excursion in 1793 by Ludwig Tieck and Wilhelm Heinrich Wackenroder to "The Gift of God" (*Die Gabe Gottes*) and other iron ore mines in Upper Franconia is widely regarded as marking the beginning of the romantic movement in Germany.[5] And so, if the geologic is the "defining strata of contemporary subjectivity," as Kathryn Yusoff writes, then the mining narratives of German romanticism could be regarded as a defining stratum.[6]

Just as mines are instrumental in the formation of media cultures, so too are they crucibles for new geological epochs. Dust collected from ice cores in the vicinity of the Quelccaya ice cap in Peru evidences "a widespread anthropogenic signal" produced by colonial mining and metallurgy in Peru and Bolivia, particularly in the fifteenth century following the rapid expansion of silver mining in Potosí after 1450.[7] The discovery of a globally legible anthropogenic signal leads the authors of that study to make two significant revisions to the prevailing narrative of the novel geological epoch informally known as the Anthropocene. Instead of linking this new epoch to the rise of atmospheric carbon dioxide levels at the start of the Industrial Revolution, they introduce the possibility of a preindustrial Anthropocene, and moreover one whose material signature consists in elevated heavy metal concentrations in the context of colonial extraction. Although the mines and caverns explored by Tieck and Wackenroder differ from those whose processing left a global trace in ice cores (in fact, the growth of Potosí may be linked to the exhaustion of silver mines in Saxony and Bohemia[8]), German romanticism is implicated in the Anthropocene in other ways. In his letter detailing his descent in 1793 into an iron ore mine, Wackenroder's description of a "new machine" to run "pumps for removing water" from the mine implicates it in what until recently was the prevailing account of the Anthropocene.[9] According to this account, posited by atmospheric chemist Paul Crutzen, the Anthropocene started "in the latter part of the eighteenth century, when analyses of air trapped in polar ice showed the beginning of growing global concentrations of carbon dioxide and methane."[10] Crutzen selects this date to coincide with James Watt's design of the steam engine in 1784, arguably the single-most important technical development behind the growing concentrations of these gases, and it is quite likely that Wackenroder encounters its predecessor, the Newcomen Engine, in the mine in 1793. This would have been just three years after Goethe was first introduced to a Newcomen steam engine, in the late summer and early autumn of 1790 during a trip to Upper Silesia, where, as Myles Jackson notes, they had been operating for several years (fig.1).[11] Goethe would go on to introduce the Newcomen engine to the mine in Ilmenau that he was overseeing at the time. This is a machine that figures even more centrally in another account of the Anthropocene, namely Gaia theorist James Lovelock's *A Rough Ride to the Future*. In reference to the Newcomen Engine and "the pressing need of mine owners with flooded mines," Lovelock argues that this need "set the start of the Anthropocene with the steam engine [i.e., the Newcomen Engine] in 1712."[12] And so it is probably not too much of an exaggeration to say that what Wackenroder and Tieck and many other romantics inadvertently witness in the mine, besides the deafening noise of these engines, is the emergence of a new geological epoch. When the eponymous protagonist of Novalis's *Heinrich von Ofterdingen* (1802) descends into the Earth to meet with a miner and describes, in the very year that Tieck writes *Rune Mountain*, how

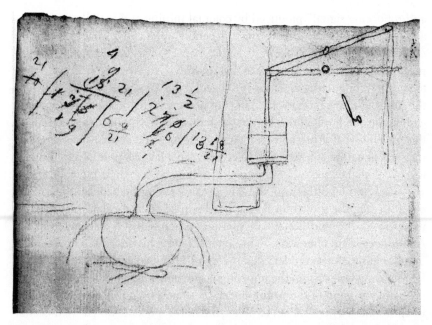

Figure 1. Sketch by Goethe of a Newcomen steam engine, ca. 1790–1791. Reproduced in Otfried Wagenbreth, *Goethe und der Ilmenauer Bergbau*. Acta humaniora, 1983, figure 15.

"tremendous births are making a stir in the depths of the Earth, distended by the inner fire of the dark womb to gigantic and immense shapes," he is also marking the monstrous natality of humanity's geologic agency.[13] If romantic narratives of descent into the mine allegorize a descent into human history, literary tradition, and sexuality, they might do the same for the ascent of the Anthropocene.

It is not so surprising, then, that some of the more prominent theorists of the Anthropocene return at crucial moments in their writing to German romanticism and its legacy, even when any link between these two periods tends to be left unexplored. At the center of *Inhuman Nature: Sociable Life on a Dynamic Planet* geographer Nigel Clark turns to the post-Enlightenment legacy of the 1755 Lisbon earthquake; in *Vibrant Matter: A Political Ecology of Things* Jane Bennett uses Kafka's tales, inheritors of the romantic *Kunstmärchen*, or artistic fairy tale, to introduce vital materialism; in *A Geology of Media* Jussi Parikka turns to Novalis's *Heinrich von Ofterdingen* to articulate an alternative deep time of the media; and most recently Thomas Ford's "Romanthropocene" evokes Friedrich Schlegel's philosophy of romantic poetry for its conception of the world as artwork.[14]

Any examination of romantic legacies in the Anthropocene, however, would do well to cast a glance at Ludwig Tieck's *Rune Mountain*, a tale of petrophilia gone awry. Particularly because romantic narratives of "enthrallment with earthly excavation" rarely "evoke the ensuing dirt and damage done by mining," as Heather Sullivan observes, Tieck's poetics of mining in *Rune Mountain*, a story of derangement and a damaged life, is all the more compelling.[15] As readers increasingly revisit romantic narratives of mining, extraction, and descent—such as Kate Rigby, who in a recent reading of *Heinrich von Ofterdingen* proposes "to exhume an ecological ethos that might provide a locus of resistance to today's political economy of extraction"[16] —it is arguably still important to seek out stories of complicity whose ecological ethos is *dark* in Timothy Morton's sense of "admitting our coexistence with toxic substances we have created and exploited."[17] Elsewhere Rigby characterizes *Rune Mountain* as a "dark rejoinder" to the utopian romanticism of *Heinrich von Ofterdingen*, and indeed the ecology of the former is nothing if not dark: underground, nocturnal, and despairing, though also vibrant.[18] However, as John Lyon points out in a recent reading of *Hymns to the Night* and *Heinrich von Ofterdingen* that draws on Sara Ahmed's *Queer Phenomenology*, Novalis dwells on the nocturnal and the subterranean precisely for their darkness and their corresponding lack of stable footing, lack of spatial orientation, and lack of temporal definition.[19] If such experiences prove to be fundamental for the establishment of a sense of orientation (or *re*orientation), then any locus of resistance to the extractivism that the mining engineer Novalis himself helped develop could be sought in moments of disorientation. The darkness and disorientation of *Rune Mountain* might be productive for an ecological ethos after all. To start with, the alluring crypticness of the nonhuman world in *Rune Mountain*, particularly where it resists the ecologically destructive anthropocentrism and anthroponarcissism inherent in prominent accounts of the Anthropocene (e.g., "the geology of mankind"), could point the way toward their revision; second, the geosocialities that emerge in the course of the narrative (e.g., "the company of feral stones"[20]) could indicate the "path of more-than-human communicative co-becoming" that Rigby suggest as a form of future coexistence to be gleaned from romantic narratives.[21]

Metal Imaginaries

Theodore Ziolkowski's "The Mine: Image of the Soul," from *German Romanticism and its Institutions*, remains the most thorough and insightful treatment of romantic mining narratives. As Ziolkowski shows, from the image of the human-mineral hybrids that haunt texts ranging from Tieck's *Rune Mountain*

(1802) to E. T. A. Hoffman's *Mines of Falun* (1818) and all the other narratives of over-imaginative young men being lured away from the flatlands to the mountains, the mine functions as a source of images for romantic visions. My impulse to reexamine the mineral imaginary of German romanticism, however, ultimately stems from a two-fold dissatisfaction with Ziolkowski's dictum that "the mine in German Romanticism is a mine of the soul, not a technological site."[22] When reviewing Ziolkowski's sources it seems abundantly apparent today that a full account of the image of the mine in German romanticism entails exploring not only the way that the mineral realm has been imagined in literature but also how the romantic imaginary is materially *from the mine*, as the Medieval Latin *minerale* ("something mined") indicates. In this case it might be more site-specific to speak of a metal imaginary, recalling along with Valerie Allen that metal derives from the Greek *metallon*, "mine" or "quarry," which indicates the relationship between metal and the act of mining.[23] Furthermore, to treat the mine as a "mine of the soul" risks overly anthropomorphizing a space whose interest for romantic tales, in *Rune Mountain* as much as any, lies in its recalcitrance to domestic concerns.

That the mine from Tieck to Novalis to Hoffman is extensively figured *in* literature as an image of the human condition does not exclude it from also serving as a technological site *for* literature. Indeed, Jussi Parikka's proposal to consider mines as sites of the emergence of technical media culture can be productively extended to encompass literary media, a prospect which Parikka proposes but does not consider at length.[24] Although the iron ore mines of Upper Franconia cannot be said to enable nineteenth-century German literature to the extent that copper and tantalum mines in the Democratic Republic of Congo materially enable contemporary digital media culture, this may be a difference of degree rather than kind. Romantic imaginaries are underpinned by material acts of extraction and they are generated in what Lewis Mumford calls the "manufactured environment" of the mine.[25] Although literary texts might be likened to an incidental mining byproduct rather than the recovered ores, this admission does not disregard the crystallization of German romanticism, in part, out of an early phase of modern mining technology. It emerges out of encounters that take place within those depths and is facilitated by acts of excavation and extraction within those depths, and in this way, it is also materially *of* those depths. Insofar as the soul is figured as possessing depth—an image greatly elaborated by the romantics—it has as its substrate the various technological achievements necessary for subsurface mining, of which the Newcomen engine is one.

A glance at the document that Ziolkowski regards as the inaugural document of German romanticism—Wackenroder's letter of June 3, 1793, detailing the initial mine descent of the period (May 22, 1793) by figures associated with that cultural movement—invites a reconsideration of the claim that the mine is an image of the soul rather than a technological site. It will be useful to read *Rune Mountain*, this fictional journey into the mountain that was written in 1802, with reference to the mining tour that Tieck and Wackenroder undertake in 1793, since both the descent into the mine and the hikes in the Fichtelgebirge serve as models for the tale.[26] The technical content of the letter can be attributed to its intended recipients: his parents, who financed the journey and expected an account of their investment. Nonetheless, the fact of the matter is that the mine in Wackenroder's letter is nothing if not an industrial and technological site. Like several of the other sites visited on the journey there are now typical "romantic" descriptions, as of the "rolling hills, gloomy, black woodlands") surrounding the Kemlas mine, but the account does not romanticize the region's industrialization. At the outset of his account of the mine, Wackenroder notes that "the Selbitz River powers an extraordinary amount of iron hammers and mills as well as a marble cutting mill for Bayreuth's marble factory" and that "one hears the noise [*das Geräusch*] of the iron hammers everywhere."[27] (Although these objects are retouched from the landscape of *Rune Mountain*, they might still be audible in the ambient noise [*Geräusch*] that opens the story.) This note is followed by a breathless inventory of the different forms of iron and stone encountered in the mine: yellow iron earth (*Eisenerde*), rough iron ore (yellow and brown varieties), glossy iron ore, iron ore occurring in globules as well as clusters, columnar iron, a variety of aragonite known as *Eisenblüte* (*flos ferri*), bloodstone, *Glaskopf* (a mineral aggregate with a glassy surface), stalactites, and so on.[28] There are observations on the working conditions and compensation of the miners, precise observation of the depth of the mine (173 feet and 4 inches), and observations of the mine's architecture.[29] The dream-like aspect of the mine, which will become the hallmark of the descents in Tieck's *Rune Mountain* and Novalis's *Heinrich von Ofterdingen*, both published in 1802, is already present when the narrow passageways prompt Wackenroder to relate a dream from childhood.[30] But in the entire account, one stretching over several pages, this reference to a dream is the only time in which the mine becomes the projection, or the screen for the projection, of a subjective interiority.

Rereading the various ascents and descents of Tieck's *Rune Mountain* against the backdrop of this descent narrative, it might become less compelling to explore how Tieck's tale places the mine in a human context—the mine of the soul—and more how it places the human in what Elizabeth Grosz calls "its properly inhuman context."[31] For Grosz this context is provided by Darwin's conception of the evolutionary descent of man and its consequences for the humanities. Although it would be obviously anachronistic to read either *Rune Mountain* or Wackenroder's letters with reference to Darwinian descent, it is nevertheless notable that these mineralogical tours unsettle the protagonists' footing on and with the nonhuman in significant ways. In this regard the most striking feature of the tour might be the startling image of Tieck and Wackenroder reduced to moving on all fours in the mine's passageway: "one can hardly walk upright in there," writes Wackenroder.[32] The physical descent into the mountain and the subsequent if momentary deviation from bipedalism kinetically rehearses an ontological kinship with the nonhuman—and in inverted form rehearses the descent of organic life from the inorganic—that is so frequently attested to by the romantics, whether in their reception of medieval philosophy or in their reception of contemporaneous scientific speculation into the history of life.

Metallurgical Flows

In *Rune Mountain*, which draws on medieval conceptions of stone's vitality and firsthand observations of the mining industry and contemporaneous geosciences alike, Tieck situates the human not only among other living things but also among nonliving things.

Accordingly, the representation of human-mineral relations in *Rune Mountain* generates questions for the humanities similar to those that Grosz poses in the wake of Darwin (for "animal" read "mineral"):

> What would a humanities, a knowledge of and for the human, look like if it placed the animal in its rightful place, not only before the human but also within and after the human? What is the trajectory of a newly considered humanities, one that seeks to know itself not in opposition to its others, the "others" of the human, but in continuity with them? What would a humanities look like that does not rely on an opposition between self and other, in which the other is always in some way associated with animality or the nonhuman?[33]

The mining narratives of German romanticism facilitate this iteration of the humanities, and in some aspects take it a step farther, by exploring various affiliations and affinities between humans and minerals and by considering not only the animality of the human, which is Grosz's focus, but also its minerality. For *Rune Mountain* the mineral does not precede but rather pervades and persists beyond the human; it is *inhuman* in the senses elaborated by Jeffrey Cohen in *Stone: An Ecology of the Inhuman*: both as "difference ('in' as a negative prefix) and intimacy ('in' as indicator of estranged interiority)."[34]

At times *Rune Mountain* will expose a continuity between the human and the mineral, between language and landform ("Rune Mountain"), and it will do so by disrupting the traditional equation of the inorganic realm with the inanimate. Ziolkowski charts how the romantics inherited the belief, reaching back to antiquity, that stones and metals grew like organic matter, as *Romanze* in Tieck's medieval drama *Kaiser Oktavianus* (1804) proclaims while airborne:

> Disclosed are the realms
> Where gold and ores grow
> Where diamonds and rubies germinate
> Quietly sprouting in the shallows.[35]

Ziolkowski traces this notion back to the Aristotelian theory of celestial influences, which was accepted by most scholastics of the Middle Ages, as well as Agricola's theory of "lapidifying juices" that were used to account for mysterious phenomena ranging from the apparent narrowing of abandoned mine shafts to the increasing density of leaden roof tiles.[36] If the influential Freiburg Mining Academy no longer regarded these influences and juices as active, its students and adherents, including Tieck's friend Heinrich Steffens and the *scientist and philosopher* Gotthilf Heinrich von Schubert, nevertheless continued to find the boundary between organic and inorganic in a state of flux. As Schubert writes in *Ansichten von der Nachtseite der Naturwissenschaft* (*Views on the dark side of the natural sciences*), "The entire kingdom of metals seems to have emerged where two worlds meet; out of the decline and decay of the inorganic, it seems to bear within itself the seeds of the new organic age."[37] Thus the *in*organic can be reassessed here along the lines of the *in*human: not what is not organic but rather what interior to, if also estranged from, the organic. Indeed, for some time now Tieck scholarship such as Sullivan's has demonstrated the liveliness of the inorganic world in the romantic imagination. In this regard it is worth remarking that Tieck's friend Heinrich Steffens, a philosopher who studied at the Freiburg Mining Academy, suggests that his "lively presentation" of the impression left on him by the sight of a particular rock (Labradorian syenite) in the Norwegian mountains directly inspired *Rune Mountain*:

> Tears poured out of my eyes; it seemed to me that the innermost part of the Earth had opened its most secret workshop for me; as if the fecund Earth, with its flowers and forests, were a chaste but light covering that hid unfathomable treasures, as if this covering were pulled back, cast away, in order to pull me down into the wondrous depths

which had opened up. The impression was a thoroughly fantastic one and it may have been a lively presentation of this impression that inspired Tieck to write his novella, *The Rune Mountain* . . . Tieck admitted that he thought of me with this novella.[38]

The geoaffect that abounds during this period, however, is multifaceted. Heather Sullivan has shown in an exemplary reading of *Rune Mountain* in the context of the contemporaneous earth sciences that the inability to maintain a stable boundary between organic and inorganic nature contributed to a widespread cultural anxiety. Sullivan reads in this anxiety a subtext regarding conflicting "ideas about the particularity of the human body." An alternative reading, however, could start from a widespread concern with the vitality of inanimate nature.[39]

Rune Mountain may facilitate calls to rethink the equation of the inorganic realm with lifelessness, yet, as this tale shows, to simply equate the inorganic with life, vitality, and animation would be equally questionable. As Elizabeth Povinelli argues in *Geontologies*, we should be suspicious of the appeal to recognize "the liveliness of the (in)animate other" as a version of the liberal appeal "to recognize the essential humanity of the other, as long as the other can express this otherness in a language that does not shatter the framework of the liberal common."[40] Here a rereading of *Rune Mountain*, in which the inanimate world is largely unrecognizable, for all of its anthropomorphic qualities, and its language inhuman, may be instructive. Not only the speaking stones but also the noisy waters and shrieking underground tremors share with the human protagonist a kind of feral common—in the words of the tale, a "society of feral stones" (*die Gesellschaft der verwilderten Steine*) (R, 202). As Alice Kuzniar shrewdly observes, the typical reading of *Rune Mountain* that focuses on the inwardness and narcissistic projection of the protagonist, Christian, "fails to take into account how Christian's imaginings assume a life of their own."[41] Kuzniar's turn of phrase is worthy of further elaboration: not only Christian's imaginings but also the minerals and mineral waters that feed those imaginings begin to take on *a* life. In drawing on a Lacanian framework that discloses the externality of the gaze, and in pointing out how that gaze is frequently attributed to rocks and metals, Kuzniar shows, as if to echo Steffens's anecdote, how "inanimate nature is possessed with a vitality surpassing human life and reducing it by comparison to irrelevance."[42]

My reading, however, is also informed by Deleuze and Guattari, whose geophilosophical writing attends to the kind of metallurgical flows that comprise *Rune Mountain*. For her part, Kuzniar's remarks on Christian's imaginings actually anticipates Jane Bennett's Deleuzian discussion of the "incipient tendencies and propensities" of inanimate things in the chapter "A Life of Metal" from *Vibrant Matter*.[43] Bennett's title refers to Deleuze's 1995 essay "Pure Immanence: A Life," which elaborates "a" life as something indefinite, incorporeal, and impersonal, a virtual life that is actualized in encounters and events and that can be extended to encompass inorganic objects. To distinguish an impersonal life from the life of a person, Deleuze observes two competing attitudes toward the dying Riderhood in Dickens's *Our Mutual Friend*: those who regard the disreputable rouge with contempt suddenly manifest great concern for signs of life when he is found dying. In the space between the life and death of Riderhood, Deleuze discerns what he refers to as *a* life: "an immanent life carrying with it the events or singularities that are merely actualized in subjects and objects."[44] As Bennett points out, a life is not only discernible in living things; drawing on Deleuze and Guattari's Nomadology plateau, Bennett argues that the complex dynamics of polycrystalline substances such as metal "best reveals this quivering effervescence" of a life and gives rise to "the prodigious idea of Nonorganic Life."[45] On the one hand, the ape-man Rotpeter delivering his vita in Kafka's "A Report for an Academy" forms the starting point for Bennett's discussion. On the other hand, Tieck's tale of whimpering mandrakes and walking, talking metals might better facilitate Deleuze's attempt to discern a life in a-subjective currents, loosened from the physiological and the organic. As Bennett observes, "metal is always metallurgical, always an alloy of the endeavors of many bodies, always something worked on by geological, biological, and often human agencies."[46] What follows are two trips through *Rune Mountain*: the first to take a listen to metallurgic flows and earth currents in all of their unearthliness and the second to take a closer look at those metals as they look back at us.

The Aberrant Earth of *Rune Mountain*

If *Rune Mountain* anticipates the continuity of human history and geologic history in the Anthropocene, it is not an Anthropocene that would encourage us to celebrate "our powers and capabilities," a typical starting point for discussions of the so-called "geology of mankind," but rather one that discloses, as Nigel Clark posits, "the position of our susceptibility to the Earth's eventfulness" and "our all-too-human exposure to forces that exceed our capacity to control or even make full sense of them."[47] This shift in perspective is enacted in *Rune Mountain*'s more-than-human story of exposure and discomposure: though ostensibly a story about an eccentric protagonist, Christian, abandoning himself to a life of extraction, it is also about the eventfulness of a landform, Rune Mountain, which serves less as a passive backdrop and more as a captivating, eccentric antagonist.[48]

From the outset *Rune Mountain* is attuned to the incipience of inanimate matter. The story opens with "waves" (*Wogen*) of the murmuring brook's "alternating melody" (*wechselnde Melodie*) that seem to speak "a thousand things in unintelligible words" (R, 181), and these have been read strictly with reference to what the narrator later calls Christian's "erratic imaginings" (*irre Vorstellungen*) (R, 191). However, the predominance of noise in the story, beginning with "the rush of waters" (*das Rauschen der Gewässer*) in the opening line, demonstrates the tale's commitment to exploring Christian's exposure to those forces that exceed his capacity to make full sense of them and its commitment to refusing to domesticate the nonhuman by presenting it in terms that most define the human, namely language and semiosis (R, 186). Attending to noise in *Rune Mountain* entails much more than attending to aural hallucinations; following Rüdiger Campe's observations on how "noise" (*Rauschen*) functions discursively as a "limit of perception," it entails attending to the limits of literature, of language, of human perception, and, correspondingly, it entails attending, through a series of liminal figures inaugurated by the unintelligible discourse of the waves, to the vibrancy of the Earth as it exceeds human and for that matter organic perception.[49] The noise of *Rune Mountain*, in a line of inquiry opened by Herder and early romantic theorists of language, perturbs anthropocentric and anthropomorphic forms of linguistic expression.[50] This immoderation is evident in paradoxical figures and situations whose unresolvable tensions structure the story: a verbal but inarticulate brook, the rune mountain, speaking stones, the earthbound Christian and his unwarranted valorization of nonprecious metals.

Things, in the tale, are more than what we make of them. Overhearing his father, a gardener, talk one day "of the underground mines and the miners" that he visited in his youth, the brooding Christian quickly develops a fascination that escalates into a resolution to leave his parents' home in the flatlands (R, 187). Having "found in a book news of the nearest large mountains," he one day makes his way for the highland and takes up hunting and forestry, the activities that define him in the opening lines (R, 188). Yet all of this is unsatisfactory, for his life does not correlate to what he had imagined it would be like. His dejection at things not measuring up, however, quickly flips into a state of captivation by things becoming much more than he could have imagined.

A turning point occurs when one day Christian absentmindedly uproots and unearths a mandrake root—*Alrunenwurzel*, an orthographic mutation of the *Alraunwurzel*, and one whose proximity to the *Runenberg* of the title symbolically registers an affinity between the vegetable, the mineral, and the semiotic. In doing so, he suddenly sets in motion "a soft whimper in the ground, which traveled underground in waves of lament, and diminished wistfully only in the far distance" (R, 188).[51] What is startling about this event, beyond the shocking sound wave and beyond the disproportionateness of the sound's duration relative to the momentary action, is how it indicates the capacity of the Earth to operate as a medium—for transmitting sound as much as recording traces. This capacity, signaled by the extraction of the mandrake root, prepares the way for understanding the Earth in *Rune Mountain* as a technological site: Christian's encounter with the mountain is mediated through a book, yet "Rune" Mountain is itself a mediated space of inscription, and in this way the tale prefigures a contemporary understanding of minerals and metals as "premediatic media material."[52]

The turning point is also one of magnitude. Several years later, while briefly reuniting with his estranged parents, Christian situates this local event in a planetary context, or, in his terms, in the context of "the whole earth." In response to his father's complaint that the stones, rocks, and cliffs of Rune Mountain have "shattered his mind" (*dein Gemüt zerüttet*), Christian responds: "No . . . I recall quite clearly that a plant first acquainted me with the misfortune of the whole earth [*das Unglück der ganzen Erde*]; only since then do I understand the sighs and laments audible in the whole of nature, if only one will listen; in plants, herbs, flowers, and trees there moves and stirs in pain one great wound; they are the corpses of earlier magnificent worlds of stone, they offer to our eye the most shocking putrefaction" (R, 202).[53]

What was initially a traumatic event for an individual telescopes outward into the scene of a planetary trauma ("one great wound"). Christian's embrace of everything inorganic and his hostility to organic life that is demonstrated in this passage and for that matter Christian's entire biography following his encounter with Rune Mountain could be dismissed as a romantic obsession; but this would reproduce an anthropocentrism that is at odds with the sentiment of the above passage as well as more recent attempts to articulate such a geotrauma. This very convergence of a planetary thought with a melancholic affect has been theorized by Nicola Merola in another context, but those remarks might help to shed light on the tale's dark ecological ethos. Merola defines melancholy as a "feedback loop that involves cultivating a radical openness to the agency of nonhumans and things and our relations with them; extending grief to encompass nonhumans and ecological processes; designing forms, objects, and rituals for enacting environmental grief; and learning to inhabit environmental melancholy as a permanent condition."[54] Christian is in this regard an exemplary melancholic subject, as evidenced by his sensitivity to the Earth's whimpering and the recognition of an impersonal agency in the mandrake and various minerals; his "melancholy" (*Wehmut*) caused by the recognition of the Earth's misfortune; his development of an intense attachment to minerals as vestiges of a lost world; and the inescapability of his all-consuming affinity for minerals (R, 192).

The planetary dysphoria that Christian experiences is first made audible by forms of environmental dissonance. In a sense the scene of uprooting condenses the entire story: a dark discovery and its dark reverberations. Following the extraction of the runic mandrake, a stranger suddenly appears, escorts Christian into the mountain, and then points him to the summit that is occupied by a mineral woman (described in the story as a *Waldweib*, or woman of the forest) whose sight exceeds his wildest dreams (and which I will discuss in the following section). The rest of the story is relatively straightforward, even when its trajectory involves a repetition compulsion whose readings are disputed: both Christian's return to the flatlands, following a loss of consciousness in the mountain, and his subsequent commitment to his recently acquired family turns out to be a temporary interval in his obsessive love of the metals and minerals to which he ultimately dedicates his life. Whether or not his attraction is read as one of inspiration or avarice, whether the tale's moral can be identified with the pieties of Christian's plant-loving and mineral-averse father, might be beside the point. It is less about maintaining binaries and more about their dissolution. Heather Sullivan's insight into the "inability of the figures to discern where one realm ends and the other begins" concerns not only the historically contested boundary between the vegetable and mineral that persisted into nineteenth-century geological debates but also how this boundary can be actively shifted and undermined by acts of extraction.[55] The tale's fundamental inconclusiveness is a reflection of its understanding of this capacity for boundary shifting. And this problem of indiscernibility is made evident in the tale's (sub)liminal noise.

Just as the Earth does not organize itself into song but rather an intermingling of nauseating noise and subterranean sound, in *Rune Mountain* the world does not organize itself into organic and inorganic objects but rather material flows that partake of both. Instead of the conventional trope of the song of the Earth, *Rune Mountain* offers us a mutant melody at the outset of the Anthropocene that could be called *the song of the Eaarth*, after environmentalist Bill McKibben's nonorthographic variant of the planet formerly known as Earth. Alongside the errancy of Christian, *Rune Mountain* explores the becoming-aberrant of the Earth: though not yet subject to the damage and transformations associated with the Anthropocene, the capacity of the Earth to take Christian by surprise demonstrates that the tale's most unsettling concern is an Earth that operates accordingly to a dimly understood geologic, an earth that responds nonlinearly to human forcings (the Earth screams merely because a plant was extracted from the ground), and an Earth whose aberrancy is even more pronounced in visual realms than sonic ones. As important as the acoustic sphere is for the tale, the textual code of visuality is supreme.

The Mineral Gaze

If the noisiness of the Earth pervades the opening scenes of the narrative, its visual excess marks the first sighting of Rune Mountain and all that follows. Kuzniar rightly reads this excess as Tieck's response to Fichte and the belief in the creativity of the romantic imagination, but there is more to it.[56] The mineral imaginary of *Rune Mountain* is also a matter of the imagination's material underpinning in the mine and thus of the fearsome capacity of the mineral to generate images. "How beautiful and alluring the old stone gazes down on us" (*wie schön und anlockend das alte Gestein zu uns herblickt*), exclaims the ominous Stranger at the outset of Christian's journey to the mountain, beckoning to the first sight of the distant landform that will spell his downfall while also indicating an inanimate world endowed with vision (R, 189). As the Stranger disappears into a mining shaft, Christian sets off for the mountain and nature becomes animated, if not anthropomorphized. He remarks, "all things seemed to beckon him forward; the stars appeared to be shining towards it; the moon cast a bright road to the ruins; light clouds rose up to them; and from the depths, the waters and rustling woods spoke to him encouragingly" (R, 190).

Just as Ziolkowksi regards the tale, for all of its eccentricities, as constituting "virtually a lexicon of romantic notions of mining," so too is its attribution of vision to the nonhuman not at all anomalous.[57] Discussing Early romantic theories of the knowledge of nature in *The Concept of Art Criticism in German Romanticism*, Walter Benjamin elaborates Novalis's argument that, in the medium of reflection, the thing and the knowing subject merge into one another on the basis of the "dependence of any knowledge of an object on the self-knowledge of that object."[58] Reception, in Benjamin's reading of Novalis, cannot be thought of as a one-sided affair; rather, it entails a reciprocal determination in which our awareness of a thing (or work of art) is dependent on that thing's own capacity for reflection and self-awareness. To illustrate his point, Benjamin cites a line by Novalis that could also be immensely fruitful for thinking through Tieck's investment in the mine: "In all predicates in which we see the fossil, it sees us"[59] This structure of reciprocity for Novalis can theoretically encompass millions of years, and for Tieck it extends to a variety of petric objects, not only petrifactions. What takes place on Rune Mountain is what Novalis would call an intensification or romanticization of reflection: the incorporation of other centers of reflection into its own self-knowledge, thereby staging not only an encounter with a mountain but a being-encountered by a mountain. The romantic vision of mines involves minerals envisioning the romantics. This reciprocal image-making is not only theoretical; just three years after and using limestone quarried just a few hundred miles away from Tieck and Wackenroder's descent into the Kemlas mine, Alois Senefelder discovers and then perfects the technique of lithography. In *Rune Mountain* Tieck offers a glimpse of a romantic lithography, but one that deserves to be thought alongside the material technique that Senefelder is developing at precisely this time.

The auspiciousness of *Rune Mountain* lies in its discovery of the technologization of perception as much as it lies in, as is traditionally held, the discovery of the unconscious. Jussi Parikka's observation that the depths of mines serve as essential places for the emergence of technical media culture has an ironic accuracy in light of the glimmering "tablet" (*Tafel*), inlaid with precious stones and metals that Christian first voyeuristically observes through the window of a marvelous palace atop Rune Mountain and then receives, through its window, from the statuesque "extraterrestrial beauty" he has been watching (R, 192). Prefiguring a touchscreen tablet device, or rather showing how a tablet imaginary exists within German romanticism, Tieck writes that the "tablet seemed to form a marvelous, incomprehensible image with its various colors and lines" and it catalyzes a remarkable metamorphosis: "An abyss [*Abgrund*] of forms and sounds, of longing and desire opened within him while throngs of winged notes of melancholy and joyful melodies coursed through his mind [*Gemüt*], which was moved to its very foundation [*Grund*]" (R, 192). The phantasmagoric experience of the tablet belies its materiality ("gleaming from the many embedded stones, rubies, diamonds, and all jewels") much in the same way that tablets in contemporary media culture are often not seen as the geological extracts that they are (R, 192). With respect to Christian it is not so much the case that the descent into the mine is a descent into the soul as it is that mining is a technical condition of the possibility of the *Gemüt*, that is, "the inner world in its totality."[60] The mineral tablet catalyzes the emergence of interiority and everything associated with it.

Just as the first glimpse of the stone tablet blinds Christian, most mineral objects encountered on the mountain are endowed with a gaze. The progression of the tale entails the degeneration of Christian's moral vision and the proliferations of a kind of mineral vision. On his second journey into the mountains, some several years later, streams ogle him like the mysterious mineral woman: "Are those flashing eyes not looking at me from the stream [*Schauen nicht aus dem Bache die blitzenden Augen nach mir her*]?" (R, 197), he asks anxiously. In one of the most uncanny passages in an already uncanny tale, after acquiring gold pieces from a mysterious stranger (and likely avatar of that original extraterrestrial beauty) in the mountains, Christian marvels at how metal "gazes at me once again" (*wie es mich jetzt wieder anblickt*) and how its "red gleam goes deep into my heart" (*daß mir der rote Glanz tief in mein Herz hinein geht!*); the sound of "this golden blood" interpellates him and interpolates itself into the sounds of music, the wind, and of speech; he remarks, "whenever the sun shines, I see only these yellow eyes, how it winks at me" (*scheint die Sonne, so sehe ich nur diese gelben Augen, wie es mir zublinzelt*)! (R, 200). It is not only the case that objects in the landscape are capable of vision; as Kuzniar observes, such objects embody "the fearsome possibility of representing to the subject more about itself than the subject ever thought imaginable."[61] The mineral imaginary, and moreover this mineral ideology of German romanticism, exposes the uncanniness of what is *mine*: an image of the mine that is not mine, the image of my soul that does not belong to me, my imaginings that take on a life, though not necessarily of my own.[62]

The close of the tale demonstrates again the extent to which romanticism is willing to immerse itself into the noise of the Eaarth, a noise that does not resolve into a signal, sounds that do not resolve into song, a planet that does not resolve into the Earth, a wanderer which does not resolve into a recognizable human figure. In his final appearance Christian is estranged from his family to the extent that he, all but faceless, is announced only as an *it*, a pronoun that was formerly reserved for stone, and here one characterized and prefixed by adjectives of intensified disintegration: *zerrissen, verbrannt, entstellt*: "It was a man in a totally tattered coat [*in einem ganz zerrissenen Rocke*], barefoot, his face burnt [*verbrannt*] dark by the sun and disfigured [*enstellt*] by a long disheveled beard" (R, 208).[63] His attempt to share his collection of "the most precious treasures" with his audience fails miserably:

> Hereupon he opened the sack and spilled out its contents; it was full of pebbles [*Kiesel*], as well as big chunks of quartz and other stones. "It's just that these jewels haven't yet been polished and filed down," he added, "that's why their look is lacking [*darum fehlt es ihnen noch an Auge und Blick*]; the outward fire with its glance [*Glanz*] is still buried in their inward hearts, but all you've got to do is beat it out of them, scare them such that their dissemblance is of no avail, then you'll see

what stuff they're made of." With these words he took a hard stone and hit it against another until red sparks emerged. "Did you see the glance [*Glanz*]?" he cried; "they're all fire and flicker [*Auge und Blick*], they lighten the darkness with their laughter, but they won't yet do it willingly." He painstakingly swept everything back into his sack and tied its cord tightly. (R, 207)

What these common stones are missing, figuratively, is their look, but literally their "eye and gaze" (*Auge und Blick*) and "glance" (*Glanz*) strictly in the English sense of a sudden movement producing a flash or gleam of light or the flash or gleam itself bot also associated with a brief look. Lacking the eyes that appear upon polishing, they fail to return the gaze of this wanderer, but in doing so they demonstrate how decentered the human has become. The extent of Christian's being beholden to the stones finally becomes apparent in the closing line: "Since then the unlucky one was never seen again" (R, 208). Upon recognizing that his stones lack a look, the hapless one promptly departs for the woods, where, upon greeting the *Waldweib*, he then vanishes, never to be seen again.

If the romantic vision of the mines furnishes an image of the human condition that has contemporary appeal, it is that of the mineralogical dimension of perception, culture, and biological life. These transformations that *Rune Mountain* documents may not yet be earth-magnitude, but in the shrieking undergrounds, lamenting plants, and wounded planet we can recognize an early glimmering of the Anthropocene imagination, where the Earth is regarded, in Rob Nixon's words, as "a hurtling hunk of rock that feels."[64] Such an image of the planet would not be alien to readers of Tieck's *Rune Mountain*, nor for that matter would it be unfamiliar of Novalis's *Heinrich von Ofterdingen* with its monstrous births deep in the Earth, nor Hoffmann's *Mines of Falun*, with its closing image of a wizened fiancée embracing her vitrified groom, unearthed after a cave-in some fifty years prior. Although the Earth in *Rune Mountain* is aberrant, the tale itself is not an aberration. As Ziolkowski rightly observes: "Tieck wrote this tale as though he were trying, at all costs, to pack in every Romantic association with mining: the mine as the locus of struggle for the pious (Christian) soul, the fatal lure of precious metals, the sexuality of the mineral realm, the lore of speaking stones, the descent into the mine as the embrace of the beloved."[65]

Tieck may have also packed in all of its blind spots, in particular the identification of women with the Earth and their subsequent marginalization, either as background domestic figure (Elizabeth) or as grotesque forest dweller (*Waldweib*). Yet the sexuality of the mineral realm is also immoderate in a way that exceeds any gender or sexual norms. It is Christian whose body dilates ("An abyss of forms and sounds, of longing and desire, opened within him") and is penetrated ("throngs of winged notes of melancholy and joyful melodies coursed through his mind [*Gemüt*]") by the mineral woman atop Rune Mountain. It is a tale both of the Earth's vulnerability ("one great wound") but also of male fragility in the face of irrepressible mineral affinities. *Rune Mountain* is less a story about coming to terms with an overactive imagination and more about coming to terms with an excessively active planet.

2
Goethe's Erratics
Wandering through Deep Time

The interest in rocks shared by several figures in Goethe's last novel goes beyond a child's amusement and a collector's enthusiasm. What it poses as *"die Neigung zum Gestein"* describes not only a "passionate interest in rocks" but also a heterogeneous and multifaceted inclination that suggest at least four variant readings within the context of Goethe's geological interests and those of his age.[1] It consists in the magnetic attraction to stone that for Goethe indicates the hangover of a primordial era in which the interaction between organic and inorganic forms was more dynamic; it consists in an intense affection for a mineral other, which is observable in the novel in a desire for intimacy with mountains that is shared by both Wilhelm's son Felix and friend Montan; it partakes in the widely held nineteenth-century belief in the tendency of the planet's climate toward a crystalline state of absolute zero, as proposed by Buffon in *The Epochs of Nature* (1778), which Goethe had read with great interest and which likely inspired his abandoned *Roman über das Weltall (Cosmic novel)* as well as many of the cold worlds of romantic literature; and it also indicates the draw of minerals in an emerging capitalist system based on resource extraction while also literalizing the reification of social relations in such a system. The popular literary motif of the cold heart, which was disseminated throughout the Age of Goethe but perhaps was most prominent in the 1827 fairy tale by Wilhelm Hauff, can been read quite effectively as an allegory of such capitalist exchange.[2]

The unsettling inclination of this novel, *Wilhelm Meisters Wanderjahre* (hereafter: *Journeyman Years*) positions it as an unlikely successor to romantic mining narratives as well as the forerunner of contemporary ecological thought characterized by, in the words of Timothy Morton, a "'humiliating' descent towards what is rather abstractly called 'the Earth.'"[3] In an age marked by a "geologic turn" in cultural studies we can observe several iterations of the *Neigung* to which Goethe attests: vibrant materialism's enthusiasm for the vitality of nonhuman bodies (Jane Bennett); a reorienting of literary studies toward the geo that has diversified into geopoetics (Schellenberger-Diederich), geophilosophy (Deleuze and Guattari; Woodard), geocriticism (Westphal) and other geomethodologies (Moraru); the constitutive role of the passions for a geophenomenology (David Woods) and the turn toward inclinations and inclined figures in recent feminist critiques of the concept of the human (Cavarero); the elaboration of queer ecologies and ecosexualities that embrace (physically and intellectually) mineral bodies (Sprinkle, Mortimer-Sandilands); the massive economic draw of mineral resources and fossil fuels in the postwar period known as the Great Acceleration, in which the exploitation of natural resources and atmospheric carbon dioxide concentrations begin to increase exponentially.[4] Thinking through the inclination of Goethe's writing and thought toward stone(s) takes on a renewed significance in the Anthropocene, which, according to one of the more prominent accounts, commenced in the latter part of the eighteenth century, when polar ice cores began to register the increasing global concentrations of greenhouse gases, and more specifically with James Watts's design of the steam engine in 1784, a date that coincides with the beginning of Goethe's decades-long study of granite.[5] Rereading the latter might stand to deepen our understanding of how the lithosphere comes to mediate culture—and vice versa—and not only in the Anthropocene.

From the two prosaic texts known as "Granite I" (1784) and "Granite II" (1785) to the 1829 version of *Journeyman Years*—the former generally considered to be the inaugural salvo of the proposed but abandoned *Cosmic Novel*— an engagement with the geologic underpinnings of culture span much of Goethe's writing career. Just as granite serves as the "firm, unshakeable bedrock" (*feste unerschütterliche Basis*) of his teacher Abraham Gottlob Werner's geognostic teaching, so too does Goethe mobilize it as the basis of geologic and cultural activity (FA, I.25:610). In the 1785 text, more commonly known as *Über den Granit* ("On Granite"), granite serves as a medium for thought, as in the case of the Egyptians, for whom "the colossal masses of this stone served to inspire the Egyptians with the idea of creating monumental works" (*Die ungeheuren Massen dieses Steines flößten Gedanken zu ungeheuren Werken den Ägyptiern ein*) (CW, 12:131).[6] Correspondingly, the layout of "On Granite" corresponds to a stratigraphic sequence, in that the text opens with the oldest stone and the most ancient times according to Goethe's reckoning, the granitic syenite of Syene, and arrives at the most recent, the lava of modern Rome, only in the final paragraph. In this way these writings demonstrate and participate in the late eighteenth-century cultural self-understanding that was mediated by a stratigraphic imaginary it had borrowed from the emerging earth sciences. Scholars, most notably Elizabeth Powers and Heather Sullivan, have probed the relationship between Goethe's poetic oeuvre and his studies of granite in the Harz mountains and elsewhere to great effect.[7] A new spatial awareness, encompassing the spatialization of time and the temporalization of nature, begins to emerge from the vertical perspective afforded by the sheer granite mountains and cliffs of the Harz Mountains, the Fichtel Mountains, and beyond. Moreover, as Andrew Piper has shown, the printed spaces of the map and the novel—in particular Goethe and Kieferstein's 1821 geological map of Germany and *Journeyman Years*—emerge as key agents in shaping readers' relationship to a stratigraphically and geologically understood planet within a larger "verticalization of culture."[8]

Poetics of the Erratic

The aim of this chapter, however, is to articulate a geological imaginary that takes shape less around vertical cliffs, stationary mountains, and imperturbable landforms and more around the errant mobility of granite within a climatologically volatile planet. Given that granite is "coded in Goethe's work as a key site to explore new forms of perception," as Piper argues, it is curious that Goethe's later and more prolific turn to displaced blocks of granite—widely known only after his lifetime as "erratic blocks" or "glacial erratics" due to their form and origin, but also as the anthropomorphized epithet "foundlings" (*Findlinge*)—has received little consideration.[9] The prominence of human foundlings in the literature of this age makes the disregard of their lithic counterparts all the more intriguing. Readers of Wordsworth, who may be familiar with the "huge stone" from "Resolution and Independence" (written in 1802, published in 1807 and revised in 1820) and the chapter-length genealogy Noah Heringman gives it in *Romantic Rocks, Aesthetic Geology*, might be surprised to learn that such huge stones are featured in Goethe's writings from 1780 to 1831. They feature most prominently in the 1829 version of *Journeyman Years*, particularly in the miners' festival (*Bergfest*) in book 2, chapter 9. Those not familiar with Goethe's writings might be even more surprised to learn that a discussion about these blocks in *Journeyman Years* is credited by later glaciologists with being one of the earliest elaborations of the glacial theory and of a former ice age. In addition to a mineral imaginary, here it might be more appropriate to speak of an "erratic imaginary."[10]

Tischbein's famous portrait of Goethe in the Roman Campagna of 1786, started one year after "On Granite" and following the "most intensive phase of his engagement with geological theories" presciently presents this inclination toward granite in this displaced form.[11] In it a partially recumbent Goethe is supported by one of several scattered blocks of granite that are reminiscent of ancient ruins as well as, if more obliquely, the image of Goethe "sitting on fallen pieces of granite" (*auf abgestürzten Granitstüken sizend*), as he notes on November 3, 1779, while overlooking the Aarve Valley in Switzerland from atop what scholars have identified as glacial erratics transported from the Mont Blanc region.[12] An earlier version of the painting has the blocks covered in hieroglyphics, thus allowing them to be identified as the imperial debris of a granite obelisk brought to Rome from Egypt by the first Roman Emperor Augustus. In the final version, however, those hieroglyphics were painted over, giving the blocks the appearance of a naturally occurring if incongruous stone that has been worked.[13] With the geologic significance of these ruins emphasized by their formal relationship to the erratic granite blocks that Goethe had been studying (and sitting on) already for several years, the painting can also be read as a portrait of Goethe's passionate interest in these orphaned stones. In addition to helping to reorient Goethe's own geological interests away from a strict focus on mountains and toward one that includes the scattered blocks, the arrangement of the objects in this painting along the horizontal axis also offers an occasion to rethink nineteenth-century studies' focus on cultures of verticality. Although Rudolf Bisanz traces the motif of Goethe's "half recumbent half seated ponderation" to the three goddesses from the east pediment of the Parthenon (ca. 438–432 BCE) and the Sleeping Ariadne sculpture (ca. 150 BCE), the image of Goethe reclining amidst a group of granite foundlings also calls forth associations with another female figure, namely the inclining maternal figures that form the basis of Adriana Cavarero's feminist critique of verticality and rectitude in *Inclinations*.[14] Cavarero's method, borrowed from Hannah Arendt, of overstating, exaggerating, and accentuating the subject's posture until it is distilled into a geometric form is also useful here. Accordingly, Goethe's somewhat awkward half reclining half sitting posture in the portrait can be taken as both a departure from the "egocentric verticality" of the masculinist mountaineer in "On Granite"—who upon summiting announces: "I stand firmly on granite" (LA, I.11:10)—and, with this lower center of gravity and reliance on the blocks for support, a movement toward a relational model of subjectivity that for Caverero is "marked by exposure, vulnerability, and dependence" to and on others.[15] What is especially notable here is that the others toward which Goethe inclines are also nonhuman.

As he observes in a letter in 1780, Goethe's outright "passion" (*Leidenschaft*) for geology marks such an inclination (FA, II.2:305). If his posture in the Tischbein portrait implies an alternative relation to the Earth than that of the domineering mountaineer, a Cavarero-inspired feminist reevaluation of his passion is further supported by the widespread designation of these objects today as "foundlings." One of the earliest publications to use the term in this context is the 1842 poem "The Marl Pit" *(Die Mergelgrube)* by Annette Droste-Hülshoff, a successor of Goethe's in many regards, particularly her lines on a geologic erratic: "Foundlings they are named, for from the breast/of the mother they are torn / Into a foreign cradle slumbering unaware, /The foreign hand lays them like a foundling child."[16] Jacob and Wilhelm Grimm credit Droste-Hülshoff with thereby initiating the geologic sense of foundling, though it is also true that the biologist Lorenz Oken had used the term in this context several years before.[17] But there is at least one instance of Goethe already in 1813 referring to a stone, though one not necessarily identifiable as a glacial erratic, as a foundling. Writing of a walk in Dresden, Goethe mentions in a letter to Christiana von Goethe on April 25, 1813: "On this same walk I bought a *foundling* [*Fündling*]. Not to frighten you that our family is going to increase, it is rather the case that, as Riemer reports, this is a strange stone, to which one cannot give a name, and which perhaps is only to be found this once [*das sich vielleicht nur einmal findet*]" (FA, II.7:211). Unlike Droste-Hülshoff's published poem, Goethe's private letter had no affect on public discourse regarding erratics, but his use of the term *foundling* for a unique displaced geologic object indexes a geosocial imaginary and a responsibility to a vulnerable mineral other that vastly predates Droste-Hülshoff's poem. Goethe's acknowledgment of a kinship with such stones, even in its disavowal, attests to an inclination, in Cavarero's sense, that carries with it a recognition of the vulnerability of the planet that will also guide this reading. It also invites a reassessment of his encounters and interactions with granite, if not also a recognition of the extent to which scholarship has overlooked these objects.[18]

Several years before the Tischbein portrait, in 1780, Goethe first acknowledges his inclination toward the particular case of the large, ostensibly displaced, granitic boulders that were known to litter parts of Thuringia, Prussia, and Mecklenburg.[19] Together with his travels through Switzerland the previous year, this forms the first of what Margit Wyder identifies as the three epochs in which he engages with these objects and the geological problems they posed: 1779–1780, 1819–1823, and 1828–1831.[20] Goethe's interest, however, is not absent in the intervening years. Throughout his travels in Germany, Switzerland, and Italy he frequently inquires into the distribution of these stones and where possible incorporates samples into his massive collection, as in the letter to Christiana von Goethe on April 25, 1813, in which he writes of having acquired a foundling. This collection grew markedly in the 1820s, starting with his receipt in 1819 of a shipment of granitic rocks gathered from around Berlin but alien to that region's lithology and presumed to be local to Scandinavia.[21] From 1820 onward Goethe regularly received information and samples of these boulders, including a curious report in 1820 from August Klaus von Preen, a landowner near the north Baltic coast, of giant ice floes littered with large granite boulders in the Baltic Sea. This report seemed to confirm the theory of J. C. W. Voigt—to whom Goethe had dedicated the 1785 essay on granite and who from 1789 to 1821 held the position of *Bergrat* (counselor of mines) at Ilmenau, whose silver mines were reopened under Goethe's direction in 1777—that the movement of these blocks could be attributed to "transport by ice" (*Eistransport*) (LA, II.8A:580–582). In a follow-up letter to von Preen of April 18, 1820, Goethe speaks of "the migration of granite blocks" and proposes the possibility of this phenomenon accounting even for those granite boulders that he had observed in Northern Thuringia. Given the proximity of Goethe's term for this geologic activity (*Wanderung*, "migration," literally "wandering") to the cultural one named in *Journeyman Years* (*Wanderjahre*, literally "years of wandering"), as well as the proximity of this letter to the novel's first edition (1821), his deep interest in these migrant rocks invites a reconsideration of the novel's epistemological horizon as well as the reception of Goethe's geology as nondynamic and static (LA, II.8A:583).

By the time of his 1829 version of the novel, granite's apparent errancy has come to present a considerable geological and poetological problem. Instead of directing his attention to the stationary granite cliffs of the Harz or the Fichtel Mountains, in these later writings Goethe focuses almost exclusively on the stationary but conspicuously displaced blocks that he knew to litter the northern flatlands of Germany. His acquisition in 1827 of specimens and a lithograph of a prominent block outside of Berlin, known as the Markgrafenstein, marks the beginning of the most intensive phase of his engagement with the blocks. As commentators have noted, his work on this geological problem runs parallel to his work on the second version of *Journeyman Years*, of whose new material at least half is completed from September 1828 to March 1829. It is likely during this period that Goethe writes the chapter on the miners' festival, which is largely devoted to the question of the blocks, and to which most readings of Goethe's literary engagement with geological problems devote the majority of their attention. No readings, however, have attempted to consider whether the concern with wandering granite blocks itself might be diffused throughout *Journeyman Years* and not just limited to the passage that discusses them.

This inclination toward the erratic and the errant can be mobilized to supplement a kinematic study of the novel, as Helmut Müller-Sievers has outlined, with a *kinetic* one folded around the movement of various forms of geologic debris. The solution to the problem of narrative motivation that notoriously plagued this novel and its readers can be explained, as Müller-Sievers argues, with reference to forms of rotational motion furnished by astronomical and kinematic figures in the novel, in particular the sprawling realm of the novel's most fascinating figure, the human and celestial body known as Makarie, as well as the many industrial machines encountered and elaborated throughout the course of Wilhelm's journey. In the face of this unrelentingly complex novel, though, Müller-Sievers also acknowledges that a fuller analysis of *Journeyman Years* would have to take into account the figure known alternatingly as Jarno/Montan and Goethe's engagement with geology. Indeed, the polarity between the terrestrial and the celestial organizes much of Goethe's thought at this time. Moreover, a study oriented toward the geologic complicates but also complements a study oriented toward the astronomical and kinematic in those very "formal aspects in which the modes of composition themselves are at issue" that Müller-Sievers studies.[22] In addition to conceiving the novel as a machine—a "*Getriebe*, a gearbox of meshing stories"[23]—attending to the geologic allows it to be conceived of as a collection of *Geschiebe* (an early name for erratics, literally "shuffled" things) in a literary "aggregate" (Goethe's own term) that constitutes something of an experimental narrative model for determining the relations between components of a heterogeneous assemblage, such as the formerly glaciated landscapes of Northern Europe or the various forms and genres (archive, aphorism, diary, lyric, narrative) aggregated in the novel.[24] A reading that attends to stones rather than stars, erratic rather than kinematic motion, blocks rather than gears, and to a narrative principle based on the (editorial and geological) shuffling of materials rather than the fate of a protagonist's drive extends the formal focus on discontinuities and inorganic development in a novel that "describes *and* abandons itself to motions that have no origins in subjectivity" while also gaining purchase on both the novel's thematic and aesthetic concern with *Entsagung* (renunciation, relinquishment).[25] The pervasiveness of errancy, a form of movement that the novel loosely links to the motif of the Wandering Jew and then disparages in favor of a vaguely Christian ideal of oriented wandering, undercuts the novel's attempts to supersede this former ideal of wandering in favor of the latter.[26] *Journeyman Years* can be read as a narrative of messy intercalation rather than straightforward supersession. And as the novel can be situated in a longer series of engagements with the problem of the erratic, I will first discuss how and why these engagements might interest us today and then how and why they figured into Goethe's geological studies—studies whose breadth is too extensive to cover in a chapter and consequently whose scope is restricted to the problem of the granite blocks.

Geologic and Epistemological Blocks

These ostensibly displaced boulders—as well as the theories of ice ages, deep time, and climate change that they helped to instigate—would go on to unsettle existing theories of the Earth arguably to a greater extent than the most widely discussed tectonic event in Goethe's lifetime, the Lisbon earthquake of 1755.[27] If this is true, then these objects might prompt us to recalibrate our readings around the erratic, the displaced, and the unconsolidated in the literature and art of this age. Accordingly, Adam Trexler's observation in *Anthropocene Fictions* that "the imaginative capacities of the novel" have been altered by climate change can be extended back to *Journeyman Years* where it contends with the susceptibility of climate to change, a possibility that the novel explores in encounters with these blocks and other markers of formerly glaciated landscapes.[28]

Trexler's argument, it is worth noting, underscores geographer Doreen Massey's observation about how the importance of such "immigrant rocks" lies not only "formal knowledge . . . but what one allows it to do to the imagination."[29] But for all that the blocks do to the imagination, it would be hard to overstate how much formal knowledge they undo. In consolidating the term for a naturalistic object, an ecology of ignorance, and the waywardness of the planet, the "block" offers insight into one of the three humiliations of ecological thinking—alongside hypocrisy and weakness—that Timothy Morton has identified in *Hyperobjects*, namely "lameness," defined as the failure of any object "to coincide with its appearance-for another object" and which becomes most salient in a time in which "humans find nonhumans pressing in on all sides."[30] At first glance, the Anthropocene diagnostic seems to imply the opposite state of affairs: humans are everywhere pressing in on nonhumans, such that many leading geologists and atmospheric chemists readily affirm their question, "Are humans now overwhelming the great forces of nature?"[31] However, Morton's hyperobjects, defined as "things that are massively distributed in time and space relative to humans" and which attempt to account for objects ranging from black holes to Styrofoam to global warming, temper human agency and scale with inhuman ones.[32] Without abandoning the Anthropocene diagnostic or denying that anthropogenic forcings significantly contribute to transformations of the Earth system, Morton's writings on so-called hyperobjects support geographer Nigel Clark's view that "whatever 'we' do ... the planet is capable of taking us by surprise."[33] One consequence of this orientation to a more-than-human scale is that, in placing the focus on the Earth's eventfulness (and secondarily our vulnerability to it), it extends lines of inquiry triggered by the Anthropocene well beyond anthropogenic events. In this context, Goethe's attraction to the strange blocks of granite evinces his openness, no matter how tentative, to the planet's inherent instability and thus to human vulnerability, even though he was working prior to any exponential uptick in greenhouse gas concentrations or even the technologies that could determine the change in those concentrations. In addition to the lines of inquiry opened by the Cavarero's feminist critique of verticality and Morton's speculative realism, those opened by the vibrant materialism articulated by Jane Bennett—developed in response to "the capacity of things—edibles, commodities, storms, metals—not only to impede or block the will and designs of humans, but also to act as quasi agents or forces with trajectories, propensities, or tendencies of their own"[34]—might help to understand the shift from Goethe's celebratory musings on granite mountains to those often tentative and unsure writings taking shape in the flatlands and in confrontation with those inscrutable blocks that turned out to be indexes of a hyperobject called the Weichsel Ice Age, as it is known in Europe, the most recent period of extensive glaciation in Europe and during which his favorite foundlings were transported to Northern Europe from Scandinavia by the movement of glaciers. The reconstruction of their journeys and, subsequently, of the climatic conditions that facilitated them unsettled existing assumptions about the Earth's history and played a significant role in determining the Earth's susceptibility to abrupt climate change, as glaciologists later confirmed. The methodology of stumbling that Goethe elaborates in the aphorism collections in *Journeyman Years* in response to these and other "blocks," whose very name registers the impairment that Morton's "lameness" registers, might then be treated as a rudimentary form of our Anthropocenic lameness, incorporated into a scientific and artistic procedure.[35]

Accordingly, the "implosion of intellectual competences" that Timothy Clark and others observe in the current period of climate change finds a predecessor in Goethe's engagement with these granite stumbling blocks, even if they act as proxies of "ordinary" rather than anthropogenic climate change.[36] (And even then it is important to recall, as glaciologist Richard Alley reminds us, that "for most of the last 100,000 years, a crazily jumping climate has been the rule, not the exception."[37]) That they were epistemological as well as material blocks is evident from the proliferation of terms in his writing: accumulations of blocks (*Blockanhäufungen*), masses of primeval stone (*Ursteinmassen*), masses of granite (*Granitmassen*), diffuse rocks (*herumliegende Felsen*), diffuse blocks of primeval mountains (*umherliegende Urgebirgsblöcke*), diffuse blocks of granite (*umherliegende Granitblöcke*), and granite boulders (*Granitgeschiebe*). The erraticism of this terminology can be attributed to the uncertainty surrounding their provenance and their inability to be integrated into a context of geological or spatial significance; moreover, the term *erratischer Block* appears in the German language only following the discussion of the *bloc erratique* in naturalist Louis Agassiz's *Études sur les Glaciers* (1840), and even then only despite the consistent translation in the 1841 German edition of *bloc erratique* as *Fündlingsblock*.[38] Although Irmgard Wagner claims that Goethe preferred the term "errant block" (*Irrblock*), a cursory search of his geological writings dealing with the granite blocks does not yield a single hit for this term, and his ambivalence around allowing errancy to terminologically and conceptually define these blocks suggests otherwise.[39]

If Goethe had ever used the term *Irrblock*, he probably would have done so in a conversation with Eckermann recorded on February 13, 1829, about Leopold von Buch and granite blocks. There, *Irrblock* would have referred less to the material block of granite and more to what historian of science Gaston Bachelard has called an epistemological block, an *obstacle épistémologique*:[40]

> Herr von Buch, he said, has published a new work, which contains its hypothesis already in the title. His writing is said to deal with blocks of granite that are found lying around here and there, but one doesn't know how and from what. But as Herr von Buch puts forth the hypothesis that such blocks were tossed forth and blasted by something violent, he indicates this already in the title, in which he already speaks of scattered [*zerstreute*] blocks of granite, and from there distraction [*Zerstreuung*] is only a step away, where the unsuspecting reader puts his head in a noose of error. (FA, II.39:307)

That the title is said to contain a hypothesis is a charge that Goethe often brings out. Nominal designations, particularly in the natural sciences, are understood by Goethe to be the abbreviated form of an often extensive act of cognition. Even if he confuses work and author in this conversation, Goethe understands the problem of scattered granite blocks. If these masses are understood as "scattered," they could be accounted for by a violent physical force, which—especially considering the vast distances over which these sometimes massive blocks were thought to have been displaced, not to mention the frequency with which they occur in Europe—would shatter his conviction that the formation of the Earth proceeded for the most part without violent upheaval and largely without mechanical forces. This was the position Goethe adopted from Abraham Gottlob Werner, the leading mineralogist of his time, and under whom Goethe, Alexander von Humboldt, and later Novalis had all studied at the Freiberg Mining Academy.[41] And so, in designating these large granite blocks not as "lying around" (*umherliegend*), which in either case is accurate enough, but with the adjective "scattered" (*zerstreut*), von Buch misleads the nascent earth scientists into thinking that they must have been physically scattered by a physical force. In reading the scattered blocks as the product of a scatterbrain, Goethe attempts to show that the designation fails to stick to the object in question—or rather that it is the designation that generates the surplus of meaning that sets the signified block in motion. If Goethe were to speak of an *Irrblock*, then, it would refer less aptly to a displaced block of granite and far more so to both the error (*Irrtum*) of conceptualizing those blocks as "scattered" and, relatedly, to the capacity of thought to be led astray by aberrant language.

The lack of closure regarding the problem of the granite blocks is compounded by the problem of an errant terminology that (mis)guides thinking. In identifying and performing in his own writing the slippage from one word (*zerstreut*) to another (*Zerstreuung*), from attribute to condition, Goethe, in the conversation of February 13, 1829, points beyond mere subjective error and to a fundamental aberrancy of the symbolic domain. It is an aberrancy that subverts Goethe's own language for the blocks, despite his efforts to secure an accurate terminology and with it not only a proper way of perceiving the blocks but moreover a containment of the planet's errant geologies. This scattering of sense is unquestionably at play and at work in an ostensibly geological matter, which, from another perspective, is also a philological one. Even Goethe's *Block*—which is the single term he consistently uses and approves of in discussions of the boulders, and whose terminological basis is derived from a consistent usage of *bloc* stretching back at least to the famous geological guide to Switzerland written by Horace Bénedict de Saussure in 1780—refers throughout its usage less and less efficaciously to a discrete mass of granitic composition. Increasingly audible is the historically obscured sense of *Block*, coterminous with the English homonym but also related to the Middle High German *bloch* and then the Old High German *biloh* or "closure, obstruction, shut place," and the Gothic *bi-lûkan*, "to close, shut" (*OED*). The block is a lock that refuses to close.

Blocked out and locked out, the erratic never secures a stabile place in Goethe's theory of the Earth, nor in his vocabulary. But the repressed returns in other iterations, in other symptomatic irruptions of a geological unconscious. *Refoulement*, the term that will later be used by French translators of Freud to render *Verdrängung* (repression), had been consistently used by geologist Élie de Beaumont (1798–1894) to describe the sudden uplift of mountains. It is invoked by Goethe in a letter written to Karl Friedrich von Klöden on the occasion of the publication of the second edition of his *Über die Gestalt und die Urgeschichte der Erde* (*On the form and history of the earth*, [1824]1829) and again in his review of Hausmann's 1827 lecture (FA, I.25:649). In the latter Goethe writes, "Instead of unthinkable upthrusting [*Aufstürzungen*] out of the deepest abysses (*Aufstürzungen*, we have to form such a peculiar word, as the French use the word *refoulement* in this situation) making demands on the imagination with such explosions" (FA, I.25:657). Although Goethe speaks of a geologic upthrusting rather than a psychologic repression he himself provides a link between the two in the moment of the unthinkable. In the announcement of the unthinkable, *refoulement* assumes both discursive registers: a geological as well as a psychological expulsion. Freud's observation that the visibility of a thing relates inversely to the degree of its repression obtains in the geological sense as well, since the *refoulement* of the object from the most abyssal depths coincides with its symptomatic appearance on the surface of the Earth. However, to speculate on the psychoanalytic provenance of this "unthinkable" is not the intention of the present work. At most it would lead past one block only to be confronted by another. My concern instead is the extent to which the erratic is on the one hand repressed as the unthinkable, rather than what this block might conceal, and on the other how the erratic can be detected, and to a certain extent treated as a figure, to gain access to crucial if overlooked aspects of his thought and writing.

In his observations tying an epistemological to a verbal obstacle, Goethe anticipates Gaston Bachelard's remarks on the primacy of the "epistemological obstacle" in *The Formation of the Scientific Mind*. As Bachelard points out, a primary modality of the epistemological block is verbal. For Bachelard the epistemological obstacle does not consist of a mental block but rather in the uninterrupted course of thinking: "It is in the very heart of the act of cognition," he writes, "that, by some kind of functional necessity, sluggishness and disturbances arise."[42] Bachelard insists that instances of inertia that crystallize in the obstacle are not to be confused with any cognitive or perceptual deficiency. Rather, thinking, simply left unchecked and unquestioned, will immobilize itself: "an epistemological obstacle will encrust any knowledge that is not questioned."[43] Left unchecked, knowledge is a condition of stasis, of walking in place, or alternatively a condition where thinking gets carried away into something that is no longer thinking, and where, in Goethe's words, "distraction is only a step away."

No statement by Goethe, however, gives more insight into the epistemological dimension of the block than his reaction to Alexander von Humboldt's *Fragmens de géologie et de climatologie asiatique* (1831), the result of an extensive scientific expedition across Asia from April to December 1829. Writing up the results of his studies, Humboldt reaches the conclusion that the Caucasus and Himalaya ranges arose out of a rift in the surface of the Earth, similar to the processes of uplift that Leopold von Buch had proposed during his own earlier travels through Norway several decades earlier. With respect to the proposed tectonic mechanism, Goethe complains in a letter to Zelter on October 5, 1831: "That the Himalayas were upraised 25,000 feet from the ground, yet loom in the skies as solidly and proudly as though nothing had happened, is beyond the limits of my head, in the gloomy regions where transubstantiation and such things dwell, and my whole cerebral system would have to be reorganized—which would really be a shame—in order to find room for such miracles" (FA, II.38: 473). A kind of transubstantiation does take place within this letter, in so far as the stark contours of the mountain merge into the stark contours of a cerebral system as inelastic as the landforms it considers. What makes this blockage curious is how unwarranted Goethe's intransigence on this issue is, how little his resistance to a dynamic Earth seems to be grounded even on his own geological principles, which understand the Earth to be in a constant state of formation, a position which, remarkably, led him to hit on the actual glacial mechanism of the blocks' transport decades before it would enjoy widespread consensus.

The block around which Goethe's theories of the Earth hover and crystallize in the 1820s, then, is doubled in the material block of granite and an epistemological obstacle posed by a sclerotic theory of the Earth that does not admit of any significant capacity for disturbance. Here, I explore how and where a concept of the erratic was blocked and where it was not, as Goethe repeatedly attempts to come to terms not only with these displaced stones but moreover with a host of political, scientific, and aesthetic developments during the opening decades of the nineteenth century. I therefore propose to install the erratic and the block as key figures of thought in Goethe's late work, principally in a reading of *Journeyman Years* and the contemporaneous geological writings. By looking at the aesthetic strategies developed in Goethe's 1828 study of a migrant stone alongside the more-than-human poetics of wandering in his 1829 novel, my aim is to show how this novel's inclination toward the erratic performs the humiliating descent that anticipates a prominent strand of contemporary ecological thought.

A Tale of Two Stones

One boulder stands out in particular in Goethe's multidecade engagement with the problem of the blocks: the Markgrafenstein (fig. 2). This massive

Figure 2. Julius Schoppe, "Depiction of the large Markgrafen Stone on the Rauen Mountains near Fürstenwalde" (Abbildung des großen Markgrafen Steins auf den Rauenschen Bergen bei Fürstenwalde), 1827. Lithograph, 42.4 × 52.5 cm. Source: Wikimedia Commons.

seven-hundred-ton granite boulder, located on the Rauen Hills outside of Berlin, is discussed at length in his 1828 essay "Granitarbeiten in Berlin" (Granite works in Berlin) published that year in his journal *Über Kunst und Altertum (On Art and Antiquity)*. The "work" that the essay describes is not restricted to the labors of the master builder Christian Gottlob Cantian and the stone masons Wimmel and Trippel, who, with the assistance of many laborers, split off a seventy-five-ton fragment of this boulder, shaped it onsite into the form of a basin, and then transported it overland and along the Spree to the *Altes Museum* in Berlin, where it would be installed in front of the museum after two years of mechanical polishing (fig. 3). (It remains there today, although the intervening years of weather and war have added to its texture.) Despite this massively monumental feat that earned the "granite basin" the popular title of the Biedermeier Wonder of the World, the work that most interests Goethe is the aesthetic work of observing the original boulder *in situ*. Not the stone that the builders took to Berlin but the one that they left behind in the forest becomes the cornerstone for a new way of thinking about

Figure 3. Johann Erdmann Hummel, "The Granite Basin in the Berlin Lustgarten" (*Die Granitschale im Berliner Lustgarten*), 1831. Oil on canvas. 66 × 89 cm. Source: Wikimedia Commons.

the Earth. Goethe's essay concerns itself less with artistic works in granite and far more with strategies of perceiving and representing those colossally inscrutable geologic anomalies referred to in the opening noun of the essay as granite boulders (*Granitgeschiebe*).

As the word *Granitgeschiebe* indicates, these "granite boulders" are neither conceived in terms of static form nor in terms of the anteriority of ruins but rather in terms of an ongoing movement of ruination; rather than as a substance, granite is presented as a thing in motion. *Geschiebe*—literally "something shoved"—typically denotes rock debris that has been "shoved forward [*fortgeschoben*] and rounded by the action of water," according to the Grimms' dictionary.[44] That the object hewn out of the Markgrafenstein took the form of a large basin has a certain irony for an object that was purportedly formed from the action of water: the theory that the seven-hundred-ton boulder reached its current location in Fürstenwalde by hydraulic force, and thus would be a *Granitgeschiebe* in the most common sense of the term, does not hold water. As we see in the essay, Goethe will attempt to play down the mobility of the boulder, yet by announcing it as a *Geschiebe* he also suggests that the landscape has been shuffled, etymologically related to *schieben*, a sense confirmed much later by the geologists that show the Markgrafenstein and its entire setting to be constituted by glacial drift.

In the second part of "Granite Works in Berlin," Goethe turns to the conspicuousness of the Markgrafenstein and its deviation—too far inland, too great an elevation—from any account that would ascribe its position to the transportation via ice floes: "It is of considerable importance that this granite rock be beheld [*vor Augen erhalten*] in its colossal setting, before it is utilized, as is happening now, for the above-mentioned project.... We would count ourselves lucky if granite here were really to be found standing in its original setting, and we would be led closer to a modest solution for a weighty geological problem that has been handled all-too tempestuously [*allzustürmisch*]" (FA, I.25:640).

In this truncated version of the essay the impulse to behold the rock, in the idiomatic expression *vor Augen erhalten*, can be seen to contain both an appeal to the Markgrafenstein as evidence and a plea to preserve it. The sense that the rock is imperiled, combined with the anticipation of it leading to the solution to a scientific problem, lends a particular weight to *erhalten* that pushes it closer to an environmentalist sense of "preservation." The notion that the milieu of a single boulder would be significant and deserving of both depiction and at least a temporary preservation, was a novel one for the time. In fact, the impulse to behold the blocks intact fueled some of the first campaigns for land preservation in Europe on noneconomic grounds. Historically, the first campaign to preserve such a boulder had its first success just ten years later, in 1838, with the *Pierre-à-bot* in Neufchâtel. In 1868 the *Call to Swiss people asking them to conserve erratic blocks*, issued by the Helvetic Society for Natural Sciences, became a popular success that led to the preservation of hundreds of blocks.[45] The case of the Markgrafenstein was different in that it was not yet evident what it stood for, yet it was widely surmised to be an important piece of evidence in the history of the Earth. That the site in Brandenburg would later be classified as a geotope, a place whose significance is first derived from the geological and second from the cultural sphere, certainly owes something to Goethe highlighting the scientific value of this object.

Goethe not only wanted to see the rock remain standing, he also wanted to give this improbable site the status of its *Urlage*, or "originary setting." In this way a "modest solution" to the "geological problem" of the boulder's apparent displacement would be secured: it would not have been violently shoved by an immense hydraulic force or by volcanic activity, as the reigning Vulcanist school's tumultuous (*allzustürmisch*) theories had proposed for similar boulders in Switzerland. A full account of this geological "problem" would lead into a lengthy discussion of the nineteenth-century Vulcanist and Neptunist schools that would only rehearse existing surveys, but these lines in "Granite Works" do much more than reanimate a dispute in the history of geology.[46] Goethe's appeal to observe the rock *in situ*, combined with his interest in the illustrations that situate it amidst its natural surroundings, also indicates an environmental aesthetics reminiscent of the pioneering work of naturalist Alexander von Humboldt in his 1807 *Essay on the Geography of Plants* as well as both his and Goethe's botanical illustrations that emphasized plants' relations to their physical surroundings. As Sabine Wilke observes of this mode of depiction, it "emphasizes the smallness of the figures vis-à-vis the vastness of nature and it turns them into insignificant elements, almost afterthoughts."[47] But whereas Humboldt was primarily interested in the relationship between plant communities and their physical surroundings, Goethe's environmental imagination here takes as its subject inorganic bodies, and moreover those that possess little cultural or economic value. Their importance is instead derived from their function as evidence and as indexes of the Earth's inhuman history. That environmental imagination and those visual depictions, where the human figure is marginalized as in the Schoppe lithograph of the Markgrafenstein (see fig. 2), also tracks with a feminist reading of Goethe's inclinations discussed earlier in this chapter.

Goethe's sensitivity to the environment here emerges in response to objects that stand in no observable connection to their physical surroundings. In drawing attention to what threatens to unsettle rather than what promises to maintain the integrity of landscape, these writings exhibit an openness to the incongruent. This incongruence extends into Goethe's own account, which oscillates between presenting the boulders as mobile (hitchhikers on ice floes) and as stationary (the adamant core of an eroded mountain). Indeed, this openness to competing accounts challenged some of Goethe's deeply held assumptions regarding the history of the Earth, yet it is an openness that was crucial for his scientific methodology, considering his known distrust of formulaic solutions, his insistence in the *Theory of Colors* on multiple experiments, and a general receptivity to difference.

Something of this attunement to the erratic can be seen in the case of the Markgrafenstein and its setting in the Rauen Hills, whose anomalous elevation is out of place in the region:

> It is situated on the left shore of the Spree River, six miles from Berlin, across from the town of Fürstenwalde, and, relative to this area, at a

high elevation of 400 feet; what is more, it is not alone and two other granite boulders can be found in the area, one already known and the other recently discovered. The summit of the Rauen Hills, roughly 300 steps north of the Markgrafenstein, rises 450 feet above sea level . . . This region is highly remarkable, in that a considerable elevation prevails here and thus seems to have caused the course of the Spree river to deviate away from the Oder River. (FA, I.25:640)

An apprehensiveness about the location of the granite is evident not only in the insistence that the setting is actually an originary setting but also in the observation regarding the anomalous elevation of the region and the apparent deviation of the Spree's course. Goethe's unease regarding the provenance of both these blocks and their surroundings turns out to have been uncannily well founded. Although the rock was blasted apart and shipped away to Berlin shortly after Goethe wrote these lines, more recently geologists have determined that this piece of fine Prussian granite in fact originated in Sweden; during the last ice age that ended some eleven thousand years ago, it was entrained by the several kilometer-thick continental ice sheet that not only picked up the 1.2 billion year-old boulder from a Scandinavian outcropping and deposited it on the Rauen Hills but also created those hills out of other foreign material deposited by the melting and retreating sheet of ice. The Markgrafenstein is still referred to as a *Granitgeschiebe* today.

To mention this twentieth-century geoscientific solution to the problem of the block's provenance is somewhat anachronistic in the context of Goethe's work, in that this exact knowledge of the Earth's history was not available to Goethe or his contemporaries. Nonetheless, his role in the study of this geological problem has been credited as being the cornerstone of the modern glacial theory and the theory of ice ages. With reference to *Journeyman Years*, the Swiss-German glaciologist Jean de Charpentier, whose father was one of the founding faculty of the Freiberg Mining Academy, directly credits Goethe with developing the theory of ice ages, prefacing his own 1841 *Essay on Glaciers* with a paragraph from Goethe's novel.[48] In order to integrate these objects into a coherent account of the history of the Earth, the primordial intactness of place that was so important for Goethe will be disrupted by theories of a more diachronic and dynamic earth, and the natural world will have to undergo a radical temporalization and historicization that Goethe could gesture at but not work out in detail. This story has been told elsewhere.[49] What has not been told in significant detail is how, to recall Jane Bennett's formulation, blocks such as the Markgrafenstein figure as quasi agents that block and impede some narrative designs while furthering others. Relatedly, in a new materialist perspective the granite basin might be understood not as a work of human hubris but rather as humans' ongoing participation in the ecstatic trajectories of inhuman things on an unthinkably volatile planet. As Jillian DeMair writes in one of the few articles to treat this subject in Goethe and Droste-Hülshoff, such a perspective "recognizes the agency of nonhuman matter and the ways in which humans are entangled as one entity among many in natural processes" and for that matter unnatural ones too.[50] As vibrant matter, the blocks that haunt Goethe's writings in their sheer persistence across decades are also haunting in a spectral sense, as relic-like revenants of an ice age—"a period of grim cold" (*ein Zeitraum grimmiger Kälte*) in the language of his last novel (FA, I.10:534)—that anticipates future glaciologists' discovery of the planet's climatic unsettledness. As a period of relative climatic stability draws to a close in the present day, climate scientists are returning to these blocks as a "climate archive" that facilitate inquiry into "the sensitivity of glaciers as climate indicators."[51] Likewise, a reconsideration of foundlings in the archive fiction of *Journeyman Years* facilitates inquiry into literature's ability to address—and be addressed by—this erratic planet.

The Years of Wandering: Of Glacial and Nonglacial Erratics

Generalizable beyond the geologic, but intimately related to it, the figure that I am calling "the erratic" crystallizes forms of movement resistant to continuity, linearity, and comprehensibility. Despite the attempt by Goethe's late writings to explain away irregularity, they nevertheless register the irregular and the erratic and even go so far as to provide a space to think through them in the erratic form, which is to say both the formlessness and the surplus of forms, of Goethe's own engagement with the problem of the blocks, an engagement scattered across genres from the epistolary to the book review, poetry, the aphorism, the maxim, dramatic works, and finally the novel *Journeyman Years*, where the erratic arguably receives the most pronounced treatment of the time, not only because this work contains the lengthiest discussions of this geological problem in any novel of its time, but moreover because the loose narrative of this perspectival novel, in accommodating so many disparate narrative interpolations (no less than six short stories written and in some cases already published decades earlier; multiple collections of aphorisms, which migrated throughout the novel in the course of its preparation for publication; and in general a profusion of settings that can be reckoned to the protagonist's inability to stay in any one place for more than three days), frustrates linear narratives and unidirectional practice of reading (constant, approving reference is made to Lawrence Sterne) and reflects on its own aberrant procedure in an ensemble of aphorisms dedicated to articulating the "course" of the natural sciences, which it defines as one of stumbling and error beset by epistemic as well as epistemological blocks. The novel is not without formal principles, such as that of the parallel stories, yet it is fascinated by the unassimilable and the irregular in a way that parallels and invites comparison to Goethe's ongoing consideration of the granite blocks and other geologic unconformities.

Just as this novel elaborates a multifaceted interest in stones, so too does it exhibit a formal inclination toward the inorganic. In an attempt to dismiss *Journeyman Years*, the critic Emil Staiger nevertheless highlights exactly those qualities that made it so alluring for modernists. He writes in 1959: "The *Wanderjahre* became a receptacle into which the poet intended to put all possible sorts of things which otherwise would have been lost among his papers or would have been scattered in unappealing individual writings. . . . Here there is no thought of organic formation [*ein organisches Gebilde*], of classical manifoldness and unity."[52] When Hermann Broch discusses the novel in his 1932 essay "James Joyce and the Present," he also identifies it with an inorganic figure, albeit more approvingly. He argues that with this novel Goethe "laid the foundation stone [*Grundstein*] of the new literature, the new novel."[53] What kind of stone was the novel for Goethe? Formally, he referred to it as an "aggregate," a mineralogical object marked by, as he writes elsewhere, "the indifference of its pieces with regard to their assemblage, their co- or subordination" (LA, I.9:203).[54] Elsewhere he refers to it as a Sisyphean stone, one that he imagined would be pushed over the crest of the mountain to menace his readers. "I will only be able to exhale once this Sisyphean stone, which so often rolled back toward me, finally tumbles down the other side of the mountain and into the public," he writes to a friend and professor of philology on January 17, 1829, adding, "Be so kind as to help me with the final push" (FA, II.10:858).

Although this novel itself was something of a minor scandal, it is itself scandalized by stones from the opening line to the closing collection of aphorisms from Makarien's Archive. The initial aphorism of that collection, one which migrated throughout the book during the editorial process, opens with an ominous stone that recalls the geological and epistemological blocks over which Goethe and his contemporaries were stumbling: "The secrets of life's pathways cannot and may not be revealed; there will be stumbling blocks [*Steine des Anstoßes*] over which every wanderer must trip. The poet, however, points the places out" (FA, I.25:646; CW, 10:417). These stumbling blocks index, among other things, the geological problem of the erratic block, which trips up the wandering Wilhelm already at the outset. Their obstinacy helps to explain how the novel struggles to present any form of gradual—in other words, nonerratic—movement. As Heidi Gideon succinctly observes, there is no wandering in *Journeyman Years*: "In den *Wanderjahren* wird nicht gewandert."[55] Though great distances are covered, though arrivals and departures are announced, the journeys themselves take up little if any space in the running commentary. Yet the awkward movement of the aphoristic wanderer, like the jerky motion of the narrative, is a form of wandering too, one that the novel encompasses in human and nonhuman dislocations alike, from the interpolated novellas of "The Foolish Pilgrim" and "The New Melusine," which follow displaced figures whose genealogies and destinations are equally inscrutable to the lithic foundlings that feature prominently in the frame narrative.

The title of the novel might lead one to think that it primarily charts the wandering of a human protagonist. It is certainly true that, from Wilhelm Meister to Faust, the protagonists of this period are the most itinerant of any in Goethe's oeuvre. Set in a "portable world" (*bewegliche Welt*) in which "the most competent man must think of himself as the most flexible [*den Beweglichsten*]" (FA, I.10:699–670; CW, 10:366–367), the novel concerns itself with a society undergoing an intense period of mobilization, as figured in the interpolated tales of foolish pilgrims and displaced laborers, prominent conversations around immigration and emigration, and the mysterious Tower Society whose watch word is "give thought to your journey [*gedenke zu wander*]" (FA, I.10:595; CW, 10:318). The problem of dwelling has to be figured anew for Wilhelm in particular, who cannot remain in any one area for more than three consecutive days. Yet it would be a false move to limit a discussion of wanderers in Goethe's writings to humans and other bipeds. The wanderer is also a more-than-human figure. In attending to the place of blocks such as the Markgrafenstein in Goethe's contemporaneous thought and writings, Wilhelm's wanderings can be recontextualized within the errancy of the more-than-human world. The itinerant protagonist presented in this novel is a *geomorphic* figure whose genealogy includes the granitic foundlings.

Already the opening lines of *Journeyman Years* lie in the shadow of the Markgrafenstein:

Overshadowed by a mighty cliff, Wilhelm was sitting at a fearsome, significant spot, where the precipitous mountain path turned a corner and began a swift descent. (CW, 10:97)

Im Schatten eines mächtigen Felsen saß Wilhelm an grauser, bedeutender Stelle, wo sich der steile Gebirgsweg um eine Ecke herum schnell nach der Tiefe wendete. (FA, I.10:263)

Most translations and commentaries render *Felsen* as "cliff" or "crag" and see this scene as articulating, through the sheer vertical perspective, a cartographic and stratigraphic imaginary.[56] Yet there is a semantic instability to *Felsen* that points to the geological uncertainty taken up in the Bergfest of book 2. If *Felsen* is construed as a "boulder" rather than a "cliff," a reading that would evade the strange topology of being simultaneously at the base of a cliff and at the *beginning* of a drop-off, then *Journeyman Years* ostensibly opens with an erratic block and thus points to an environmental imagination oriented not toward fixed stratifications but rather toward the plotting of an erratic mobility.

Although the opening scene is a commonplace of Romantic literature, the intactness of this topos is undermined on a number of levels. The physical setting has become unstable, the ground has become precipitous, and the word for "setting" is itself unsettled, as seen in the gliding of *Stelle* into *steile*, "spot" into "steep," in the opening sentence. These various instabilities signal an environmental imagination in which setting, in the words of Lawrence Buell , is no longer "ancillary to the main event" but is itself eventful.[57] The profusion of settings cannot be solely reckoned to the protagonist's inability to stay in any one place for more than three days; the ground underfoot is in motion. Thus, the years of wandering begin at a standstill with a sedentary wanderer sitting on a steep path onto which all sense of movement has been displaced. That the steep path turns reflexively and in an inverted word order (*wendet sich*) is not only a turn of phrase; in the subsequent mountaintop scenes it is suggested that the ground is in motion: "A mountain formation of medium height seemed to strive [*schien heranzustreben*] upward, but never approached their own elevation. Further along the land grew flatter [*verflachte es sich*] but there too, curious formations [*seltsam vorspringende Gestalten*] jutted up [*zeigten sich*]" (FA, I.10:289; CW, 10:114). There the erratics again intrude into the landscape as "curious formations" and with them the Earth takes on a strange reflexivity that further facilitates the projection of the nonhuman setting into the foreground.

The earth is no longer a mere setting or backdrop but rather, in the words of Bruno Latour, an "actant," defined as "any entity that modifies another entity" in a variety of settings from the lab to the novel.[58] In the penultimate scene of the novel, just prior to Felix's fall, the unsettling movement of erosion and displacement again accords a reflexivity to the Earth: "The boat was gliding down the river beneath the hot midday sun . . . wheat fields grew near [*näherte sich*] the stream, and a good soil came so close that swirling waters, throwing themselves [*sich hinwerfend*] up at spots [*Stelle*], had mightily attached the soft earth and swept it away [*sich gebildet*], so that steep cliffs of a considerable height had formed" (FA, I.10:744; CW, 10:416).

The presentation of the elements as active agents is not at all a capricious one. Goethe ascribes a "wild and brutal way" to the four elements in his 1825 essay *"Versuch einer Witterungslehre"* ("Toward a Theory of Weather") (FA, I.7:397; CW, 12:147); *Faust II* also attempts to account for the "aimless strength of elemental forces" (FA, I.7:397). More specifically, the writing of the penultimate scene of *Journeyman Years* cited earlier can be accounted for by Goethe's travels to and studies of the geologically and geothermally active town of Karlsbad in the years 1807 and 1808. These studies overlap with his four-month stay in 1808, from May 28 to September 7, during which he wrote five novellas, four of which would be interpolated into *Journeyman Years*, and the fifth would be *Elective Affinities*. During this period, he also publishes a geological commentary on a series of minerals which he and the lithographer Joseph Müller collected from the region ("Joseph Müllersche Sammlung"). In this commentary, which coincides with the conception and to a large extent the writing of the first edition of *Journeyman Years*, the Earth is the scene of an elemental restiveness that will later become prominent through the same rhetoric in *Journeyman Years*, as seen in the passages cited above. As Wolf von Engelhardt notes of the Karlsbad commentary, the sentences in which stones have as predicates either reflexive or transitive verbs are twice as frequent as predicates that express a static state of affairs.[59] Stones and rock formations present themselves, express themselves, transform themselves, let themselves be seen, produce themselves, spread themselves out, alter themselves, and they also conceal.[60] In this drama of things, mineral agents take humans as accusative objects: they direct our attention, they address us, they come together to make formations. And these mineral actants are some of the most significant instigators of Goethe's thought in this period. What Heather Sullivan writes about mountains in *Faust* is perhaps even more true for the foundlings of *Journeyman Years*: "they actively destabilize his—and our—assumptions about 'passive matter' and recontextualize human endeavors in their physical environment."[61]

Earth often appears in the novel not in the form of imperturbable monoliths but rather as debris in transit. The first station of the novel, set in a ruined monastery, follow Wilhelm and his son Felix, who has acquired a "passionate interest in rocks" (*Neigung zum Gestein*) during the opening trip through the mountains (FA, I.10:287). When the latter stumbles across "stones . . . left over from a large quantity which a stranger had recently shipped from here" in the monastery's chapel, Wilhelm and Felix abruptly break off the planned trajectory of their journey (FA, I.10:287; CW, 10:113). That stranger turns out to be an old acquaintance of Wilhelm—Montan, the one from the mountains—who in mobilizing this stone mass participates in the elemental restiveness of the Earth encountered elsewhere in the novel.

These nonglacial erratics, although they are set in motion by the work of a human hand, lead to the one who presents the opposite state of affairs, namely Montan, the human animated by the mineral world. Elemental forces of attraction guide Felix and Wilhelm to this secluded geologist: first through an attraction to the remnant of a shipment of rocks, they are then "drawn in" (*angezogen*) by a patch of gentians; then, following the reverberations of a hammer striking rock, Wilhelm finally reaches the summit after Montan "pulled him up" (*zog ihn aufwärts*) (FA, I.10: 291; CW 10: 114). Their reencounter takes place on a granite summit that bears a striking resemblance to the setting of "On Granite," Goethe's ecstatic, prosaic ode to the Harz mountains from 1785. "You should note," Montan instructs Wilhelm and Felix, "that you are now sitting on the oldest mountain formation, the oldest rock in the world" (FA, I.10:287; CW, 10:114). But the similarities between the 1785 and 1829 texts cease there. Montan concludes their conversation a few pages later: "these [lifeless cliffs or stark boulders] at least are not to be comprehended" (*diese [starre Felsen] sind nicht zu begreifen*) (FA, I.10:291; CW, 10:116). This attitude presents a striking contrast to "On Granite," in which extremely robust lines of communication had been established between the organic and inorganic world, and where Goethe in 1785 could write of an inorganic nature "vast and eloquent with its still voice" (FA, I.25:314; CW 12:132). Already in the 1821 edition of the novel, though, granite has become obdurate and withdrawn from its cultural enframing, no longer a good conversation partner and no longer capable of supporting the cosmic reveries of the years 1784 and 1785. And so, in the 1821 version of the novel the discussion with Montan breaks off with his promise to plunge so deeply into the Earth that he will never be found again, in order to lead a conversation, but one that is "mute and unfathomable" (*ein stummes, unergründliches Gespräch*) (FA, I.10:53). He is never heard from again in that edition of the novel. But in the space of the unfathomable and ungrounded—that of the erratic, that of deep time—novel conversations and novel conversation partners are emerging.

Montan's reticence—particularly notable given that he is the character in the novel most capable of providing an account of the Earth—lends a geological context to the novel's subtitle, *The Renunciants* (*Die Entsagenden*). While this subtitle casts the human characters as renunciants, for Goethe the term *Entsagung* has an epistemological significance. *Entsagung*, according to Goethe biographer Nicholas Boyle, represents "the renunciation of the hope of possessing the object."[62] The blocks that crop up throughout in the novel are for Goethe and Montan the exemplary object that confronts you in its strange otherness and that is withdrawn from conceptual understanding. In the context of Montan's geological renunciation, the novel could be said to explore what Lawrence Buell terms an "aesthetic of relinquishment."[63] Of note here is not only the form of relinquishment by Wilhelm that involves foregoing a fixed dwelling, physical contact to loved ones, and material possessions; moreover and more radically, the novel relinquishes a subject function, which is redistributed among the various characters and interpolated novellas, and it also relinquishes of an image of the Earth as a passive setting or static backdrop.

Goethe's study of the Markgrafenstein in 1828 would have offered an impetus to resume this conversation. And so, in the 1829 version of the novel Wilhelm and Montan reunite at the occasion of the Bergfest in book 2, though this too becomes the scene of a blockage. As Wilhelm attempts to elicit a response to the question, posed already in book 1, regarding the meaning and scope of Montan's mining and mountaineering activities, Montan responds only with his signature *Keineswegs!* "No way!" he says, "The mountains are silent masters and they train silent pupils" (FA, I.25:533; CW, 10:278). As they reach the top of the mountain and the highpoint of the festival, a remarkable "conversation proper to the place" (*ortgemäßes Gespräch*) ensues: a lofty conversation on a mountain, with a mountain (Montan), and celebrating mountains and metal deposits of that particular region. But the site-specific conversation soon turns away from the mountain and toward what is from the mountain: to the nonlocal, the exolithic, what could be called *ortungemäß*. In the brief period before its violent interruption, this conversation suited to the site of the mountain shifts away from the ground on which they stand and toward the granite blocks in the flatlands, toward the geological controversies Goethe was participating in at the time, and toward the origins of the planet's geologic formations.

In this heated argument practically every major early nineteenth-century theory of the Earth is represented. Adherents of the so-called Neptunist school, following a position held by Werner, argue for the precipitation of the Earth out of a primordial ocean (hence the epithet "Neptunist") and cite as evidence the fossilized remains of sea animals on the high mountains; they are followed by the Vulcanist school, which proposes the formation of the surface of the Earth out of volcanic activity; many of the guests, however, are unsatisfied with both positions and so an alternative is proposed, an exaggerated form of Leopold von Buch's theory of uplift, in which preformed mountains are ejected from the interior of the Earth, and finally, a more obscure theory of mountain formation is proposed, attributable to Johann Ludwig von Heim in which entire mountain chains fall like meteorites from the sky.

One group in particular directs the attention of the conversation toward anomalous "larger and smaller masses of rock found strewn [*gestreut*] over many lands" (FA, I.10:531; CW, 10:279). Like Goethe's study of the Markgrafenstein, contradictory positions are floated but not resolved, but unlike the essay form the novel allows these positions to be distributed to different voices without the need for a dominant position. Although the novel seems to frown on the inconclusiveness of the conversation and the volatility of certain conversers, at the same time both novel and novelist have seized on the loose narrative form of the novel to render a planet marked by this very volatility and open-endedness. Perhaps most notable—in the context of the contemporary confrontation with abrupt and irreversible climate change—is that the proposal of former glacial age can be found in a novel published nearly a decade before Karl Schimper coins the term *Eiszeit* ("ice age"). Just prior to the mountain conversation is cut short, a few quiet guests make this radical proposal:

> Finally two or three quiet guests invoked a period of fierce cold [*einen Zeitraum grimmiger Kälte*], when glaciers descended from the highest mountain ranges far into the land, forming in effect slides for ponderous masses of primeval rock, which were propelled farther and farther over the glassy track. In the subsequent period of thaw, these rocks had sunk deep into the ground, to remain forever locked in alien territory. In addition, the transport of huge blocks of stone from the north might have been made possible by moving ice floes. (FA, I.10:531; CW, 10:279)

Although these lines have typically been read as milestones in the discovery of former ice ages—as mentioned earlier, pioneering glaciologist Charpentier goes so far as to quote them as the epigraph to his landmark study—perhaps the allure of these lines, and the granite objects that inspired them, has more to do today with their intimation of the planet's climatic volatility. What is at stake is not only the reconstruction of the planet's history but also the intimation of its unstable climatic futures.

These early proponents of an ice age encounter a block, not in the stone but in the resilient beliefs of their interlocutors: "However, the somewhat cool views of these good people did not make much headway. The general opinion was that is was far more natural [*naturgemäßer*] to have the world be created with colossal crashes and upheavals, wild raging and fierce catapulting" (FA, I.10:534; CW, 10:279). Those proposing a catastrophic origin of the world—whether tectonically or volcanically—intervene, and, under the influence of wine, the festival nearly concludes with deadly brawls. This abrupt ending, combined with the inconclusive conversation, leads to the collapse of the world for the protagonist: "the well-ordered, fertile, and populated world seemed to collapse into chaos before his mind's eye [*vor seiner Einbildungskraft chaotisch zusammenzustürzen*]." (FA, I.10:534; CW, 10:280).

To be Continued: Steps Toward a Glacial Theory

Although the conversation in *Journeyman Years* abruptly breaks off, and the world breaks down, it continues elsewhere, just as the last words of the novel enigmatically promise: "(to be continued)" (FA, I.10:774). As Andrew Piper argues in his reading of the novel and its intertextuality, "the printed book was reconfigured in the *Wanderjahre* as an endlessly regenerating system of texts" and this is also observable in the mountain festival's conversation.[64] It is continued in Goethe's notebooks, in letters, book reviews, and in sketches of further scenes not included in the novel. Among the latter is an untitled fragment containing a conversation between several unnamed speakers about large blocks of granite scattered throughout the plains of Northern Germany. Entitled *"Erratische Blöcke"* ("Erratic blocks") by the editors of the 1894 Weimar edition of Goethe's works,[65] where it was first published, renamed "Gespräch über die Bewegung von Granitblöcken durch Gletscher" ("Conversation on the movement of granite blocks via glaciers") in the Frankfurt edition (FA, I.25:646–647), no doubt because of the anachronism of the Weimar title, and appearing under *both* titles in the Leopoldina edition (LA, I.2:377–378; LA, I.11:309–310), the problem of what to name this unnamed fragment reiterates uncertainties regarding its subject matter. Already that the conversation in *Journeyman Years* is taken up elsewhere—in a discarded text—is part of a poetics of the erratic. This draft of a conversation about blocks is itself a kind of foundling, without a fixed name, without any identifiable speakers, and without any apparent discursive context. This apparent lack of context, which reiterates its object of inquiry, make this text particularly interesting. Even a cursory reading of this fragment confirms Wolf von Engelhardt's assumption that the conversation consists of a sketch for the "Bergfest" in *Journeyman Years*, but Engelhardt does not examine the reasons for its noninclusion (FA, I.25:1272–1273).

Much of this fragment maps onto the one in the published novel. Just as the dialogue there shifted from the alpine blocks of Switzerland to those of northern German flatlands—"the transport of huge blocks of rock from the North" (FA, I.25:534; CW, 10:279)—in the corresponding passage in the conversation fragment, the blocks now found in Northern Germany are also the result of ice floes from the North. (FA, I.25:647). The conversation continues, albeit by another speaker barring the southbound path. In this way a proposal found nowhere in the novel emerges, in a discussion regarding an elevated coastal area that gives the town of Heiligendamm its name (now known to be the product of glacial deposits):

> No way; Northern Germany had its own granite rocks, though weathered, they sank down and lie in scoured sand; the Heilige Damm originates as far up as the Norwegian skerries, and it might also be the case that the ice detaches much of it and transports it to the south. One cannot make me believe that the stones found lying around the Oderbruch region, that the Markgrafenstein at Fürstenwalde has come from abroad; they came to rest in situ, as the remains of large masses of rock that decayed and fell in on itself. (FA, I.25:647)

Keineswegs, no way: in this word the basic feature of the erratic is announced, namely, the nonevident course of its movement. The voice of Montan also announces itself, the one whose statements are in most every instance prefaced with this signature phrase. But in the fragmentary conversation, *keineswegs* signals more than a speech impediment. It signifies the truncation of the distance between the provenance of the stone and its current setting. "No way" is reiterated in the absence of a path, the absence of distance between the mountain and the stone. In this case there is no erratic. In place of the migrant stones the speaker claims to discover the remnants of local granite. Here the Markgrafenstein that was the focus of "Granite Works in Berlin" reemerges and with it the argument that these immobile stones disintegrated in place, thereby leaving the integrity of the site undisturbed.

Yet in the final lines the fragment switches course yet again, concluding with a reference to the very ice-raft hypothesis that was rejected in the preceding paragraph. *Doch,* the speaker begins, making reference to the granite-laden ice floes on the Baltic coast that were documented by Voigt and von Preen: "But I do not want to reject a course from the North; still to this day large masses of ice trek across the sound, carrying pieces of granite that they have plucked off and transported. The toll collectors of Göteburg should confirm these sightings, in order that we might arrive at concepts more in line with nature" (FA, I.25:647). Unfortunately, however, the toll collectors were never summoned.

Only a few writings on geology follow this one, and they block this line of inquiry. Goethe instead returns to the hypothesis of *Urgebirg* or "primeval mountain" anywhere erratics are found, a hypothesis that admittedly had little support in the scientific community of the 1820s. The Urgebirg takes shape in two closely related texts, *Umherliegender Granit* ("Diffuse blocks of granite") (1829) and again in *Geologische Probleme* ("Geological problems") (1831). A subtle immobilization of the block is contained already in the title of the 1829 text (literally: granite lying around), which deals with those masses of blocks (*Blockanhäufungen*) found in Northern Germany that are said to provide certain evidence not just of a period of glaciation but more importantly of an ancient mountain range in Northern Germany. "Diffuse blocks of granite can be variously inferred" (*Umherliegende Granite können verschiedentlich abgeleitet werden*), Goethe writes, whereby the connotation of "eliminated" in *abgeleitet* also becomes audible: "1. As remains of rocky cliffs that stood in the same place [*Stelle*] and that, for the most part eroded, left their most solid remnants behind . . . like the Landgrafenstein rock and certainly significant others that remain to be discovered" (FA, I.25:644).

These descriptions of the ground conceptualize it as an invariably fixed site (*Stelle*). Standing or lying in place, this rubble is further distantiated from the mobile disposition inscribed in the erratic: replaced *in situ*, immobile rocks secure a stabile relation between the surface outcrop and the underlying matrix.[66] But the proliferation of these texts and the remarkable provisionality of their hypothesis suggest that the conceptualization of the blocks as remnants must not be glossed simply as "remains" of an original but also "remainders" left over and unaccounted for by any comprehensive theory of the Earth.

Two years after "Diffuse Blocks of Granite" and following a visit by Alexander von Humboldt, who gives a brief account from his tour through Russia, Goethe returns to the *Urgebirg* as a solution to the problem of erratic formations in Northern Europe. He begins this account in "Geological Problems" by addressing the contemporary observation of granite-bearing ice floes, but he definitively announces that this cause is only secondary, the erosion of a local mountain being the primary cause (or effect) of the blocks: "While we recognize stone types of the Nordic regions in Northern Germany, it does not yet follow that they come from there" (FA, I.25:656). Instead of reinstating the granite blocks to an existing mountain range far north of Germany, Goethe establishes a mountain range in place of the remaining block. He closes this fragment by characteristically valorizing the stationary: those who treat the blocks as erratics, and in doing so overturn some of the fundamental associations of the time, do so only for the sake of an explanation: "The fantastic ways of those who want explanations! What is solid and unshakeable is said to grow and move itself, what eternally moves itself and changes is said to be stationary and at rest, and all of this merely so that something can be said" (FA, I.25:656).

And so, in this rehabilitation of the unshakeable, the erratic is eliminated. But in the process of yet again explaining away the irregular, the incalculable, and the inexplicable, another proposal and another text emerged. In this way the proliferation of writings, even where they attempt to downplay the erratic behavior of these stones, delineate the outline of an erratic—an inconsistency, a block—as their very basis. They are too inconclusive and messy for any straightforward history of geology, and so the present chapter has not sought to furnish such a history or to integrate Goethe's various studies of these objects into any of the heroic narratives of the golden age of geology. In Goethe's studies, the erratic takes shape where it is not named as such, where it resists integration into an existing discourse or a fixed theory. Only as such do they stand to inform a practice of reading, which, taking stock of the displaced and dislocated, takes up the errata, the paralipomena, the unexplained, the unimaginable, the unfinished, the repressed, and the marginalized as integral if unintegrated sites demanding attention, critical intervention, and new strategies of perception.

Previous readings of the Wilhelm Meister novels have noted how the cultural activity of wandering, for Goethe, becomes increasingly fraught with anxieties in a historical moment marked by rapidly increasing mobility. Without discounting political and sociological accounts of this anxiety, the present chapter has attempted to show the role that geological anxieties play in shifting attitudes toward and depictions of wandering. Not unlike the novel's Tower Society, which is on guard against excessive wandering, Goethe's function in the geological writings surrounding *Journeyman Years* is primarily one of containment. Similarly, not unlike the inability of the novel to avoid overextending itself, the geological writings, in their unsuccessful attempts to contain the erratic blocks, begin to articulate a discourse of the erratic by their very performance of an erratic discourse. Similarly, Goethe's theory of these objects is itself glacial, in the banal sense of proceeding (and receding) gradually over a protracted period of time, and in this way unlike the sudden flash of his other naturalist discoveries. However, just as "glacial" today is no longer a dead metaphor for gradual movement, but is instead, as Rob Nixon writes, "an iconic image of unacceptably fast loss," so too do Goethe's scattered writings on granite take on an unexpected and unanticipated sense: as forms of engagement with the volatility of the planet's climates.[67] Although this form of loss is most associated with the visual sphere, for example in the case of the time lapse photography of receding glaciers in the film *Chasing Ice* (2012), the inclination of recent climate change fiction toward the experimental novel form, as in Liane Dirk's *False Heaven* (2006), Christian Kracht's *Methane* (2007), and Ilija Trojanow's *Lamentations of Zeno* (2011), continues on that erratic trajectory of ruptured forms initiated by Goethe in his last novel.[68]

As one of the several planetary transformations that the humanities and the literary arts attempt to contend with today, climate change came to be understood through the study of erratic blocks from Central Park to Brandenburg and beyond, objects that played an important role in the reconstruction of the Earth's climatic instability in the Pleistocene epoch and before. As Dipesh Chakrabarty has argued, the planetary crisis of climate change has profoundly shaken up disciplinary distinctions and dogmas; what is required, he writes, is to "bring together intellectual formations that are somewhat in tension with each other: the planetary and the global; deep and recorded histories; species thinking and critiques of capital."[69] Goethe's overlapping studies in geology and meteorology, and the articulation and development of some of these theories in literary prose, suggest that it is not only a case of bringing together seemingly incompatible intellectual formations but also bringing them *back*. If his geological theories themselves proved to be erratic blocks that are largely passed over in histories of geology, Goethe's way of pursuing these geological problems seems startlingly prescient.

3
Many Stranded Stones
Stifter's Spectral Landscapes

In an experiment that measures the viscosity of pitch, a glass-like substance brittle enough to be shattered by a hammer, a funnel of this material has been suspended since 1927 over a small glass jar at Queensland University. Initially heated and poured into a sealed funnel, the sample of pitch was allowed to harden over a period of three years before the stem of the funnel was cut.[1] In the time that has elapsed since then, this mass of pitch has been pouring through the funnel at the rate of roughly one drop per decade. Over the course of the experiment, which is projected to run for at least another one hundred years, a total of nine drops of pitch have fallen from the funnel into the jar. Other experiments elsewhere register a movement even more glacial: since 1917 a mass of pitch has moved six millimeters in a thin eighty-millimeter-long funnel stem located in Wales, at a rate calculated at one drop per thirteen hundred years. Knowing these properties, Kelvin used this material to construct artificial glaciers in the nineteenth century.

The extremely protracted movement of pitch at indoor ambient temperatures also makes it an appropriate literary model for the Austrian realist writer Adalbert Stifter, whose novella *Die Pechbrenner* (The pitch burners), written in the winter of 1847 and published in 1848, crystallized over the next four years into *Granit* (*Granite*), one of the better-known novellas in his 1853 novella collection *Bunte Steine* (*Many-Colored Stones*). Twentieth-century readers of Stifter, from Walter Benjamin to Georg Lukács to W. G. Sebald, have criticized his writing's affinity to the painterly *nature morte*, where "time, the measure of all things . . . seems not to pass."[2] Even Stifter's friend and noted geologist and climatologist Friedrich Simony likened his writing to a "petty detail painting of unimportant things."[3] Instead of the vibrant life of a narrated world, critics argue, Stifter's ostentatiously excessive description of everyday minutiae offers a petrified reality devoid of the capacity to depict social change or any form of shock. In this regard the revision of *The Pitch Burners* into *Granite* could be read as an allegory of these reifying tendencies: a world animated by the traffic of pitch-peddlers is superseded by a lifeless boulder. But just as the previous chapter saw the granite block as a site to explore the novel perception of the Earth as intemperate, in Stifter it can be seen as a site to explore perception of the Earth as unreliable. Even in the revised form of *Granite* the instability of pitch underpins the story, and it does so as a dialectical image, in motion even when at an apparent standstill. A reading of *Granite* calibrated to flows and movement of the lithosphere, no matter how intermittent and protracted, might register an unsettling mobility and vibrancy of Stifterian landscapes where others have only observed their reification and mortification. Instead of reading the confrontation with stone as an escape into a "salubrious stillness" (*heilsame Starre*), the geological formations of these narrated worlds pose a menacing moment, one in which stone projects into and perturbs the narrative.[4] Sam Frederick's assertion that Stifter's lengthy passages of description and discussion in the novel *Der Nachsommer* (*Indian Summer*) "disperse its potential for plotting, dissipating its eventfulness almost entropically so that any conventional narrative progress slows to a pace where time seems no longer in control" is perhaps not only a function of a thirteen-hundred-page "narrative seeking to escape from temporality" per se but also a function of a narrative, like *The Pitch Burners*, *Granite*, and other novellas, that seeks to register and describe other, more-than-human temporalities.[5]

What some readers disregard as petty, uneventful, boring, and glacially paced has been recognized by others as a pointed inquiry, characteristic of literary realism, into "modes of perception and representation" of what Elisabeth Strowick, after Erich Auerbach, calls "dull duration."[6] Drawing on Goethe's apprehension of the atmosphere as an aggregation of phenomena, Strowick reads the aggregate structure of Stifter's atmospheres—and the corresponding aggregate structure of perception—as a primary scene of engagement with the problem of representing temporality and duration for a realism of the nuance. As productive as a reading organized around Stifter's description of atmosphere proves to be, one organized around the medium of the lithosphere affords another opportunity, if one beset with even greater challenges, to observe modes of perception and representation of duration. The dispersed plotting and dissipated eventfulness of his narratives, particularly where they deal with the appearance of the lithosphere, can be read as a form of engagement with the problem of reading and perceiving at earth magnitude and moreover on a far more dynamic planet than had been thought, as

Stifter's engagement with developments in the natural sciences suggests. Images of stone-rich landscapes in his stories are typically subject to ongoing alteration in ways that correspond to the epistemological status of the Earth in climatological inquiry around the time of their composition, as Deborah Coen's magisterial *Climate in Motion: Science, Empire, and the Problem of Scale* testifies. As has been documented in *Climate and Motion* and elsewhere, Stifter's writing and thought is informed by a number of sources, including extensive contact with his physics professor at the University of Vienna, Andreas von Baumgartner; demonstrated familiarity with Baumgartner's prominent book on natural history; a likely familiarity with Charles Lyell's actualist geology from *Principles of Geology*, not to mention Goethe's geological writings and thought; as well as his friendship with the geologist and climatologist Friedrich Simony, starting with their first meeting in 1845.[7] Simony's account of his three-day winter expedition to Dachstein Mountain in 1842 informs the story *Bergkristall (Rock Crystal)* from *Many-Colored Stones*, just as Simony himself figures as the model for the geologically inclined protagonist of *Indian Summer*.[8] Although these relationships have been traced out in detail elsewhere, they have primarily been explored in relation to *Indian Summer*; moreover, Stifter has gained the reputation of "the man responsible for embedding the Earth sciences in Austria's literary canon," what is less well understood, though, is how *Many-Colored Stones* establishes parallel lines of inquiry to, rather than merely embedding findings from, the emerging earth and climate sciences.[9]

Dynamic climatology's challenge of conceptualizing, to quote Deborah Coen, "interactions across scales of space and time, from the human to the planetary" is also taken up in the realist program that Stifter elaborates in the preface to the 1853 novella collection *Many-Colored Stones*.[10] In a notable passage regarding geomagnetic storms and electromagnetic waves, Stifter demonstrates a subtle receptivity to planetary phenomena. As Deborah Coen notes, it was during the same years as the composition of *Many-Colored Stones* that Karl Kreil, the founding director of the Central Institute for Meteorology and Geomagnetism (Zentralanstalt für Meteorologie und Geomagnetismus, later Geodynamik), was leading "the project of measuring magnetic and meteorological variations across the surface of the Habsburg lands."[11] What Stifter writes is that the actions of a man who strictly observes and records the minor deviations of a magnetic needle may seem trivial; however, when we consider that these measurements are performed simultaneously around the world, and that these minor deviations occur simultaneously across the planet, we can perceive "that a magnetic storm passes over the whole earth [*die ganze Erde*], that the entire surface of the Earth simultaneously feels, so to speak, a magnetic shudder."[12] These lines invite us to reread Stifter's "detail painting of unimportant things" as potential glimpses into remarkable events unfolding at earth magnitude or across deep time. The magnetic storm offers an example of a significant geophysical event that is initially only registered partially and as a thing (a deviant compass needle); as such it fails to fit in any paradigmatic opposition between the large and the small while offering a model that calls into question Georg Lukács's criticism that "individual parts no longer carry the weight of the concrete moments of the plot" in Stifter's description-rich writing.[13] The case of the deviant needle also offers an instance of a "shock confrontation with marginal events" that Martin Swales as characteristic of the novella form in German-language literature: "I would argue that the mainspring of much novelle writing is the contact between an ordered and reliably interpreted human universe on the one hand and an experience or set of experiences that would appear to conflict utterly with any notion of order or manageable interpretation on the other."[14] In the novellas collected in *Many-Colored Stones*—and here I focus on *Granite* for reasons that will be apparent—the geophysical realm furnishes a number of those marginal events and experiences. Keeping in mind the image of the surface of the planet *feeling a shudder* running through it we might better appreciate their attempts, in those scenes that stage a confrontation with earthy objects, to expand the coordinates of the "experientially perceived reality" (*erfahrungsmäßig erkannte Wirklichkeit*) that for Friedrich Theodor Vischer is the foundation of realist literature.[15]

The Vibrant Earth of *Many-Colored Stones*

A number of readings of Stifter over the past decade—most notably those of Peter Schnyder, Sabine Schneider, Timothy Attanucci, Lindsey J. Brandt, Elisabeth Strowick, and Tove Holmes—have challenged earlier generations of readers who regarded his landscapes and their description as static and inert.[16] In one of the more prominent rereadings of Stifter, Schnyder calls attention to a dynamism of the Earth and its narrativization in the form of what he calls the "shock" (*Erschütterung*) of a poet-geologist protagonist suddenly coming to terms with the deep time of an alpine landscape in *Indian Summer*.[17] Similarly, W. G. Sebald indicates this dynamism when he writes of Stifter's "shocking materialism" (*erschütternder Materialismus*) in light of the erosion of a metaphysical order in his earthly and earthy stories.[18] Incidentally, the "shocking materialism" that Sebald observes in Stifter's presentation of things can also be understood in terms of the materialism that Jane Bennett articulates as "vibrant materiality." Bennett's "vibrancy" resonates with what Stifter's critics call an *Erschütterung*, which is often translated as "shock" but literally describes the "shaking" or "shattering" associated with seismic disturbances, as in the *Erderschütterung*, or "earthquake." In the absence of narratives of seismic activity, but acknowledging the vibrancy of the Earth in Stifter's stories, the current chapter extends recent readings of the inhuman in Stifter with Bennett's admonition in *Vibrant Matter* to explore "the capacity of things . . . not only to impede or block the will and designs of humans but also to act as quasi agents or forces with trajectories, propensities, or tendencies of all their own."[19] The vibrancy of the stones in a collection of stories called *Many-Colored Stones* is also signaled in their very polyvalence, and it is sustained in their capacity to introduce those "shock confrontations" in which Swales identifies a main feature of the novella form.

However, the particular form of shock in which the vibrant materiality of the ground might be registered exposes a significant tension in readings of Stifter. For one of his most prominent readers, Stifter's writing is characterized by an *inability* to articulate the shock that Sebald and in particular Schynder attribute to the experience of alpine landscapes. As Walter Benjamin writes in 1918, "The ability to somehow present 'shock' ['*Erschütterung*'], whose expression man seeks primarily in language, is absolutely lacking for him" (GS, 2.2:609). Why? Briefly, as I will take this up again in chapter 4, since Benjamin locates the capacity for shock (*Erschütterung*), like revelation (*Offenbarung*), in the metaphysical-acoustic sphere of the word, and since he identifies the absence of movement and sound (*Ruhe*) as the "basic feature" of Stifter's writing, then his writing is marked by an inability to present shock.[20] At first glance Benjamin's observation has traction even in those alpine scenes of *Indian Summer* that Schnyder identifies as shocking. As much as the novel elaborates the protagonist's awe at geologically active landscapes, and much as those scenes disrupt the narrative with a cascade of questions and inquiries into the deep history of the Earth, in the end the passage through the mountains would seem to remain little more than an inconsequential interpolation in the career of the protagonist and in the course of a novel whose tranquility is relatively uninterrupted by the excursion into the tumultuous history of the Earth's formation. Be this as it may, Benjamin's argument rests on shaky ground. In the first place, and as those and other passages attest, stillness is not reliable as a basic feature of Stifter's writing. Moreover, it is important to recall that for Benjamin the question of how to present shock, whose expression, after all, man seeks in language, has no straightforward answer. Already the expression "to somehow present 'shock' [*irgendwie '*Erschütterung*' darzustellen*]" indicates a blockage in the attempt to articulate a specific technique for its depiction, and by further placing "*Erschütterung*" in quotation marks, he exposes a certain uneasiness that extends beyond his powers of articulation. It is a problem that he will return to years later when he applies the term to Brecht's poetry and points out that "the word *Erschütterung* ['shock'] contains the word *schütter* ['thin,' 'sparse']."[21] A new look at shock, then, is made possible by noting and exploiting the undefined behavior and presentation of shock in Benjamin's critique.

If the *Erschütterung* is not one that can be so easily articulated, then perhaps the vibrancy of and in Stifter's texts could be sought in what remains unarticulated in his writing: for example, both in the *Schutt* or "debris" that marks his geologically active landscapes and in the gaps and minor inconsistencies in the narrative produced by his notorious revisionary practices that can be partially reconstructed by those archived remainders, known in Stifter philology as "discarded pages" (*abgelegte Blätter*), of published versions (*HKG*, 2.3:25). Like these discards, the apparently superfluous stones and sparse spaces in his narratives cannot be dismissed as insignificant. Readers of Stifter have demonstrated a propensity to pore over the colorful and classifiable stones featured in the titles of *Many-Colored Stones* with the venerating gaze of an antiquarian. In contrast, this chapter critically considers less congenial blocks, detritus, gaps, and heaps of debris scattered throughout those stories. This concords with readings of Stifter that are attuned to erosion, as a challenge to human perception and empirical science that his stories seek to address. It also concords with Jane Bennett's location of a vibrancy and vital materialism in the unconsolidated structure of polycrystalline materials—in the dynamic behavior of loose atoms within the vacancies, imperfections, and defects in those structures. Stifter, too, was drawn to a form of vibrancy in the unfathomable granularity of geologic materials. As he writes in the opening line of his autobiography, "The smallest grain of sand is a wonder that we cannot fathom."[22] The unconsolidated behavior of *Granite* recalls, at different levels of granularity, the vibrant materiality of polycrystalline solids, and it also indexes social upheaval and epistemological disturbances. In some cases the geologic objects in *Granite* bear the legible traces of a violent history, but in other cases it is the sketchy spots and apparent vacancies in the landscape—what is too noisy to serve as data for the geologist—that encrypt these shocking experiences and planetary shudders.

Many Stranded Stones, Many-Stranded Stones

The inclination toward the erratic and erosive in *Granite* can be situated within a wider "orphaned imagination" of the nineteenth century that manifests in Stifter's stories with orphans such as the "brown girl" in *Mica*

(*Kazensilber*), eccentrics including the mute daughter in *Turmaline* (*Turmalin*) and the reclusive pastor (himself the descendant of a Swedish foundling) in *Limestone* (*Kalkstein*), Mignonesque figures in *Forest Spring* (*Waldbrunnen*), and beyond.[23] Although the narrator of *Granite* is not a foundling per se, his biological parents are strikingly absent from the narrative. The mother appears at the outset of the story to beat her son and then retreats, only to return at the end to silently offer him reconciliation while he was drifting off to sleep, and the father is only briefly glimpsed while lying down in a sepulchral light at the close of the story. However, it would not be accurate to orient Stifter's imagination solely around human orphans. Stifter's interest in genealogy is tied up with his interest in geology: like Goethe and Droste-Hülshoff, Stifter is drawn to the problem of the foundling as a social and geologic phenomenon.

The exemplary scene of an encounter with a kind of foundling is found in *Indian Summer*, in the opening chapter ("Broadening the Horizon") of the second book, which precipitates the one Peter Schnyder refers to as a shocking experience. As this scene looms so large in Stifter scholarship, it will be instructive to consider it before turning to a more protracted engagement with the problem of a foundling in the earlier novella, *Granite*. Halting before a pile of stones at the side of a mountain road, one spring the protagonist Heinrich Drendorf encounters an accidental cairn both marking an impasse in the mountain pass and bringing the narrative to a halt. Unlike the stone sculptures that captivate Drendorf in the house of his mentor, Freiherr von Risach, this heap of stone produces an obstructive rather than an instructive experience.

> In the spring, when I had left our capital and was walking slowly up a grade behind the coach, I once stopped by a pile of stones [*Haufen von Geschiebe*] they had taken from a riverbed and strewn [*aufgeschüttet*] by the road. I regarded this with a sense of awe. In the red, white gray, dark yellow, and speckled stones all with flat rounded shapes, I recognized the harbingers of our mountains; I could tell the craggy home of each one, where it had been separated and from whence it had been sent out. Here it lay among comrades whose birthplaces were often many miles distant from its own; all of them had assumed the same shape, and all were waiting to be broken apart and used on the road. (*HKG*, 4.2:27–28; *IS*, 190)

Encounters with a variety of stranded objects will be reiterated over the following pages, and one could trace in this encounter a link between the narrator's sudden investment of attention in a debris pile and an uneasiness before the daunting pedagogical project of this bildungsroman, namely the ability, in the words of Eric Blackhall "to give expression to the 'whole' which binds together the parts of a landscape and gives them meaning."[24] To give expression to this whole, however, entails glossing over those unassimilable foreign bodies that Stifter can neither gloss over nor adequately gloss and which suddenly appear in his literary landscapes in the form of aggregates and erratics.

As was the case for Goethe, the description of these stones as *Geschiebe* ("shoved" or "shuffled" stones ranging in size from pebbles to boulders, of glacial or fluvial origin) exemplifies the complexities of description of lithic objects on a geologically dynamic planet.[25] Crystallized into a noun substantive, *Geschiebe* preserve in their name a sense of movement within a static object, yet these stones come to a rest only in passing. As such these objects pose an aesthetic challenge, and one that Stifter takes up in another medium concurrent to writing the above passage. *Movement II (Bewegung II)*, an 1858 painting of a river rock from a series of sketches called *Movement, flowing water (Bewegung, strömendes Wasser)* that he first mentions in 1854, offers another scene to think through the problem of representing motion in static objects (fig. 4). *Movement II* depicts a small river stone, which could be classified as a *Geschiebe*, submerged in what appears to be a calm and crystal-clear brook. The absence of visibly flowing water in the painting, however, means that the movement implied in the title appears to be a misnomer. Christian Mösender asserts that "movement" refers to this "potentially dangerous, but now peaceful water," which is to say, it refers to a point in time not depicted in the painting when the flow rate is higher; Lindsay Brant suggests that movement refers to the rock, as evident from its abraded surface and visible subjection to processes of erosion.[26] As prominently as the phenomenon of erosion figures elsewhere in Stifter's stories and geological imagination, I would suggest that the depiction of movement to which Stifter directs the viewer's eye is less a function of the deterioration of the stone, its declensional *ab*rasion, and more its ongoing transportation by turbulent fluvial processes, as evidenced in its flat, rounded form and as suggested by the proximity of this painting to the composition of the above-cited passage about river rocks from *Indian Summer*.

The painting's extraction of this single stone from a larger riparian scene rehearses the industrial extraction of the stones from the riverbed in the scene from *Indian Summer*. The aesthetically extracted and industrially extracted stones stand out and present a problem different than that confronted by Stifter's predecessor Goethe: the movement of the latter is interrupted by their displacement, from riverbed to roadbed, that disrupts and circumvents the

Figure 4. Adalbert Stifter, *Die Bewegung II*, ca. 1858. Oil on canvas, 24 × 32,4 cm. Courtesy of Zeno.org.

natural-historical logic of succession and stratification. Scattered (*aufgeschüttet*) on the roadside, their movement now describes a destratification and an errancy neither confined to nor determined by the course of the river. Nor does this passage describe a distanciation that would preserve an original site from which distance could be measured, but far more a distanciation from the possibility of geographically localizing either origin or destination. In the ensuing scene of recognition, the matrix of the stones is remembered only on the verge of its dismemberment and scattering. The desire to recognize their origin, which entails the imaginary restoration of these orphans to their craggy homes, and the imaginary restitution of a shattered unity, is frustrated by their utilization for the material basis of an expanding transportation network in a country about to undergo an explosion of urban migration and a massive dislocation and restructuring of geological formations. The moment of extraction and the subsequent diversion introduces a fold into the otherwise uninterrupted "way things go" in Stifter, irrevocably disrupting the unfolding of natural-historical processes.

This doubly "marginal event," in terms of its roadside setting and apparent insignificance in the greater scheme of this bildungsroman, nevertheless also constitutes a major shock confrontation in the novel. The shock is not so much registered in any positive expression as much as it is in the perturbation of the narrative. In the line immediately following the encounter with the rubble, Stifter writes: "In particular the thoughts came to me [*Besonders kamen mir die Gedanken*]: why is all that here, how did it come to be, what is it made of and what does it tell us?" (*HKG*, 4.2:266; *IS*, 190). The syntax of this sentence registers the marginalization of the subject and his displacement in the form of a direct object to the periphery of thinking. Although Stifter—a naturalist with a training not so dissimilar from the one that attributes to his protagonist—has enough of a background in geology to understand these blocks in terms of geohistorical processes, the vast timescale necessary to present the static blocks as markers of a dynamic geomorphological process explodes the human scale of the narrative, and disrupts the plot with an episodic series of loosely coordinated impressions, observations, descriptions, and a several page-long litany of questions regarding the mountainous region that is his destination after leaving the capital. "Will a good deal, will everything completely change again?" the novel asks, and each question affirms the previous: "In what rapid sequence will it proceed? If through the influence of wind and water, the mountains are constantly broken off, if the rubble falls down, if they are further split and the river ultimately brings them to the lowlands in the form of sand and stone [*Geschiebe*], how far will this go? Has it been going on for ages? Immeasurable layers of stone [*Unermessliche Schichten von Geschieben*] in level areas confirm this assumption. Will it continue for ages?" (*HKG*, 4.2:268–269; *IS*, 192). The historical time of the novel is greatly expanded, such that human history becomes—in the words that close this breathless pages-long meditation on geology that proceeds from the roadside scene with the extracted river stones—a mere *Einschiebsel*: an afterthought, an aside, an interpolation that has linguistic kinship with the *Geschiebe*:

> If any history is worth pondering, worth investigating, it is the history of the Earth, the most promising, the most stimulating history there is, a history where man is only an interpolation [*Einschiebsel*], who knows how small a one, and can be superseded by other histories of perhaps higher beings. The Earth itself preserves the sources of this history in its innermost parts just as in a room for records, sources inscribed in perhaps millions of documents; it is only a matter of our learning to read and not falsify them by eagerness or obstinacy. (*IS*, 192)

Moreover, these stones offer an alternative trajectory to that of sedimentation and stratification, one which has been scarcely considered by Stifter's readers. The interpolation of human history into the history of the Earth would serve the function of smoothing the former of its irregularities, inconsistencies, and revolutions that troubled Stifter, were the linear trajectory promised by natural history not itself complicated by the heap of river stones that initiated this sequence of thought.

Arguably the most elaborate engagement with the problem of displaced stone as it indexes sociological and geological disturbances can be found years before *Indian Summer* in *Granite* and specifically the imposing rock whose description occupies the opening will be the principal focus of the rest of this chapter. Already the title *Granite* is a kind of foundling; no object in the story is every said to be composed of granite, and the word *granite* does not appear once in the entire text even though a large stone is one of the most pronounced actants in the story. Although none of the eponymous stones appear in their respective stories in *Many-Colored Stones*, with the exception of *Limestone (Kalkstein)*, it is also the case that in no other story is a single stone as prominent as in *Granite*, which makes the nonappearance of granite that much more conspicuous. In calling attention in the opening lines to the lack of any recorded memory of how the block came to occupy its site, *Granite* calls attention to these absent origins and these spectral inscriptions:

> In front of the house in which my father was born, close to the entrance, lies a large eight-cornered stone in the form of a very elongated cube. Its side surfaces are rough-hewn, but its top surface has become so fine and smooth from much sitting as though it had been covered with the most artful glaze. The stone is very old and no one can recall having heard of a time when it had been set there. The most ancient old men of our house had sat on the stone just as did those who had passed away in tender youth and who slumber beside all the others in the churchyard. (*HKG*, 2.2:23; *Gr*, 7)[27]

In its resistance to being inscribed in any oral or written history and moreover in its resistance to being ascribed to any human activity, this stone embodies what Eric Downing identifies as the two bad-faith fictions of all realism: "that its proffered world is uniform, stable, and unchanging; and that it is untouched by human agency, is immanent, pure reality."[28] That this fiction is in bad faith can be seen in the example of a stone that has evidently been shaped by human agency (the rough-hewn edges, the slab of sandstone that serves as a base) and yet whose placement seems to precede the work of any human agent. Situated on the verge of the domestic sphere, its basis lies beyond any human reckoning or other attempt at domestication.

In the indeterminability of its provenance, and its relative inability to be meaningfully integrated into the story it frames, this block appears *stranded* in the politically and historically charged sense that Eric Santner imbues to it in *Stranded Objects*. Santner focuses on those objects stranded by the Holocaust, objects that contain the remains of a cultural inheritance and are fragmented and poisoned by an unspeakable horror.[29] In the case of the community living in the wake of the devastating, depopulating plague in *Granite*, this block of stone is a kind of stranded object. In its muteness and peculiar texturelessness in a story that struggles to commemorate a forgotten history, the block becomes an emblem of this unspeakable distant past. It is also an emblem of an unspeakable recent past: the storyline of *Granite* can be read as an object stranded by the popular uprisings of 1848, events that reframe the traditional, ritual-based society of *The Pitch Burners* as an idyllic and thoroughly anachronistic one. Stifter further strands this society by rewriting it as *Granite* and retouching the more disturbing elements. As we will see, the three prominent stones that function as narrative landmarks in *The Pitch Burners*, and whose relations show that the traumatic past can spill over into the present, are much more isolated in *Granite*, owing to the omission of an important storyline. Like the pieces of quarried marble that fascinate the narrator, the idyllic elements of the setting of *Granite* appear as "remnants of an old fallen world" (*Überreste einer alten untergegangenen* Welt) in post-1848 Austria (*HKG*, 4.2:40; *IS*, 26).

The complicated texture and textuality that I will trace in the story through a series of close readings also makes another meaning of "stranded" audible. If the proliferation of blocks, fragments, and orphans suggests a "stranded imagination" of the idyllic *Many-Colored Stones*, those stones are also intensely *stranded* in the sense "of thread, rope, or similar: arranged in single thin lengths twisted together" (*OED*). What *Bunte Steine* signals, beyond color, is the textual quality of being *bound* and thus the entwinement of the textile, the textual, the literary and the lithic. *Bunt* ("many-colored") is related to *binden* ("to bind"), such that the expression *bunt* took its significance from the inevitably variegated colors in ligatures, bands, and other garment-like strips of fabric and garments. Although an etymological link between *bunt* and *binden* is suspect—not the activity of binding but the appearance of bound fabrics seems to have lent *bunt* the association with colored—the title for this collection, *Bunte Steine*, marks how stone is bound up with other materials and also how humans are bound up with stone and the earth.[30]

The robust association between the bound and the *bunt* is evoked throughout *Many-Colored Stones*, not just in *Granite*. Motifs of linen and limestone are extensively interwoven in *Limestone* (*Kalkstein*), the novella that follows *Granite* in the collection. *Limestone* is as fascinated by the threadbare undergarments of a pastor (the poor benefactor of the earlier German version, *Der arme Wohltäter*) as it is by the region's eroded limestone geology, particularly as they both manifest an intense weathering and wornness.[31] Grey hues predominate in a landscape that is less appropriate for a collection of many-colored stones and more for a collection of many-stranded ones. Moreover, the "significant folds" (*deutliche Falten*) in face of the pastor are bound together with the fractured landscape, whose "denuded" (*nackt*) appearance likewise resembles the pastor who has "not a single hair on him," a correspondence that introduces intimate connectedness between the human and inhuman in the novella (*HKG*, 2.2:65).

This peculiar nexus of textiles, mineral textures, and literary texts extends even into the reception of Stifter in the twentieth century, particularly in Peter Handke's *The Lesson of Mont Sainte-Victoire* (*Die Lehre der Sainte-Victoire*), a text organized, like *Granite*, around a walk and the problem of how to narrate a many-hued mountain, famously painted by Cezanne, and everything that hangs together with it. The solution for Handke's narrator, an avid reader of Stifter, to "the problem of the transition" is supplied in equal measure by the tectonic geology of this mountain and a brocade, satin, and damask coat sown by the narrator's trekking companion and clothing designer Domenika Kaesdorf: the latter solves the problem in the process of fabricating a coat of these heterogeneous materials, whereas the former solves the same problem through contemplative reflection on, and narration of, "a fault between two strata of different kinds of rock" and within it a "point" where "one stratum penetrates most deeply in to the other."[32] Only through interspersed reflections on both the textilic and the stratigraphic "problem of connections and transition" in a text which begins with and is pervaded by observations on Stifter's *Many-Colored Stones* is Handke able to reach the truth of storytelling, which, he writes in a vocabulary reminiscent of Stifter's reflections on realism, "is perceptible only as a gentle something in the transitions between sentences" (*nur an den Übergängen der Sätze als etwas Sanftes zu spüren*).[33]

Unraveling *Granite* (Many Spectral Stones)

Although the importance of the linen industry in Stifter's Austrian hometown of Oberplan, coupled with the biographical fact that his father was a weaver before he became a linen merchant, stands to inform a reading of *Many-Colored Stones*, scenes of unweaving offer a better paradigm for understanding his textual practices. What makes these writings into texts in an emphatic sense is not only the complexity of their textural wovenness but moreover a textile-like capacity to be shredded and recombined into another form. In this regard, Sebald's comparison of Stifter to a priest—not the one celebrating the liturgy of an absolute order, as Emil Staiger wrote, but rather the one who tears his robes to shreds in Kafka's *A Country Doctor*—especially suits *The Pitch Burners*.[34] Out of fabrics torn into strips, new lines and lineages are always emerging: "He took out the knife, which he always carried with him, then he took off his jacket, and he began to cut a strip out of it, then another, and then more. He bound [*band*] these strips together, so that the first was bound to the second, the second to the third, and carried on until he had a long linen chain that reached to the Earth. 'Now you can tie [*anbinden*] a bread roll to it,' he called below" (*HKG*, 2.1:46).

Started in 1847, completed in 1848, and published in 1849, *The Pitch Burners* (*Die Pechbrenner*) contains a number of significant passages and developments that were deleted in the course of its revision for its publication as *Granite* in 1853. The grounds for these alterations and revisions have been extensively considered and tend to point to Stifter's conservative responses to the popular uprising of 1848.[35] An alternative account might find in the sheer reliability of Stifter's storylines—their capacity to be unbound and rebound, like Joseph's jacket, and like the narrative of *The Pitch Burners*—a narrative unreliability that would problematize the scheme of order-disruption-reinstated-order according to which his stories are often read. And this intimately related to the problem of an unreliable ground in *Granite*. The image of a child stranded on a massive boulder known as the Hutfels and tearing his jacket into shreds, included in *The Pitch Burners* but not in *Granite*, furnishes a suitable leaping off point for this reading.

In both *Granite* and *The Pitch Burners* stones project into the narrative and menace the young protagonists. *The Pitch Burners* features three significant stones: the massive Hutfels, on which the child named Joseph was once outcast; the smaller but more prominent block in front of the house (*Gassenstein*), on which the narrator's feet are smeared with viscous pitch by one of Joseph's descendants and on which the narrator and grandfather sit as the latter relates the story of the Hutfels; and an even smaller but similarly shaped stone (*Stein des Vorhauses*), located in the vestibule of the narrator's childhood house, on which the narrator is deposited by his mother after being beaten by her for having tracked pitch throughout the parlor. These stones are the anchoring points of the story and the history it narrates. The link between past and present—between the plague's devastation of Joseph's entire family due to his carelessness and the narrator's accidental soiling of his parent's house, between the orphan Joseph and the narrator cast out by his parents—is mediated in and by these stones.

Both *The Pitch Burners* and *Granite* tell the history of the Oberplan region through a narration of events surrounding a past outbreak of the plague. This history is pointed out by the grandfather of the narrator while they walk throughout the countryside one day in the narrator's distant childhood. The grandfather's journey is a routine one for him; but one day his young grandson is allowed to accompany him for the first time, as he has violated the order of the family home by tracking pitch throughout the house. Punished and cast out, the grandson spends the day with his grandfather on foot, learning about the region on a lengthy interpretive walk, while the mother and cleaning staff furiously attempt to scrub the floors clean. Their walk concludes where the story opened, on top of that prominent stone between house and street.

In the course of its revision into *Granite*, the story of *The Pitch Burners* is itself scrubbed of unsuitably violent material for the intended juvenile audience. To that end the entire Hutfels scene of Joseph's abandonment is discarded, along with everything except for a passing reference to this rock, which haunts the constellation of the other two stones that remain narratively significant in *Granite*. Thus the continued interest of *Granite* for a more mature audience is that it not only tells a memorable story replete with remarkable descriptions of landscape but also that it inevitably narrates its own acts of revision and erasure along the way.[36] In one of these scrubbed passages, partially cited above, a young Joseph has been abandoned by his father on the massive rock known as the Hutfels, accessible only by a ladder that has been removed, for having invited a family suffering from the plague into their house. While Joseph's parents and nearly the entire population succumb to the plague and therefore are no longer in a position to assist his descent, the outcast son survives his ordeal only by tearing his jacket into strips and binding these together into a line, onto which another recently orphaned child name Magdalena, who hears his calls for help, first attaches a sturdier cord and then food, a water jug, and blankets. This enables him to survive until a ladder can be procured. The boy's solution to a situation in which his continued existence is threatened corresponds to the procedure of a writer whose writing fails to cease only by shredding and rebinding existing texts. Indeed, most of the material of *Many-Colored Stones*, with the exception of the story *Mica* (*Katzensilber*), had been published under different names and other forms in journal editions, but underwent extensive reworking for its republication.

The linen chain lowered from the rock, while facilitating Joseph's survival, also literalizes the complexity of history, genealogy, and narrative in a story that proceeds by unbinding and rebinding. Linearity and circularity are the predominant patterns of movement throughout Stifter, and they are no less contradictory here than elsewhere.[37] *Granite* and *The Pitch Burners* unfold while the narrator and his grandfather set out from their home on a circular itinerary that covers long since established lines of commerce and communication. They do so while narrating the towns and other sites and places that have crystallized across great swathes of time out of the movement of goods and people. Although Joseph's linen lifeline in *The Pitch Burners* prefigures the continuity of lines and in particular that of lineage (in both versions Joseph and Magdalena later reunite and reproduce), the passage describing the linen chain and moreover the entire narrative around the Hutfels is notably absent from *Granite*. In its place is an apparently simpler storyline: Joseph discovers and comes to the aid of a fatigued Magdalena and not the other way around. The Hutfels is all but completely excised, receiving only a brief mention as the site of the ancestral home of the pitch burners, but not otherwise integrated into the narrative.

The apparent superfluity of the Hutfels seems to confirm the observation by Martin and Erika Swales that "everything in this story is so organized as to intimate continuity beyond the catastrophic, stability beyond the destructive moment."[38] After all, even with the circumvention of the Hutfels in *Granite*, Joseph still meets Magdalena, years after the end of the plague they manage to reunite, and thus the lineage from the outcast Joseph to the pitch peddler Andreas, who paints the young narrator's feet and causes him to be cast out of his own home for a day, and thereby triggers the grandfather's narration of this genealogical and regional history, is upheld. Yet the absence of the Hutfels in *Granite* is not without consequences for the fabric of the story. In *The Pitch Burners*, the Hutfels is the primary object linking a past traumatic event with its contemporary reverberation—even the "glaze" (*Glasur*) of the prominent stone whose description opens each story and on which the narrator has hit feet befouled by Joseph's descendent recalls the "smooth" (*glatt*) surface of the Hutfels on which Joseph was abandoned (*HKG*, 2.1:45)—and so the absence of the Hutfels in *Granite* undermines the significance of the opening stone (*Gassenstein*) and everything that unfolds around it. The diminished prominence of the Hutfels also suggests that these landscapes are less stable and reliable than we have been given to believe. A reading sensitive to what has been repressed, as well as to the related problem of a writing that cannot impress itself enduringly, exposes the ongoing erosion that is also the story of *Granite*.

The attempt to scrape together the missing links needed to give the landscape in *Granite* more coherence inevitably extends the continuation of the practices of unbinding and rebinding that characterizes Stifter's own textual practices. The philological scrutiny that has been applied to Stifter's revisionary practices, here and elsewhere, inevitably resolves the text into *scruta*: "old or broken stuff, trash, frippery, trumpery" (*OED*). To scrutinize means to search even to the rags, a practice that here yields the phantom of the boy tearing his jacket into strips to bind together and lower himself to the Earth. In this way the many stranded stones could be said to be *spectral stones*, both in terms of their colorfulness and in their ways of haunting the narrative. In the light (not) cast by the spectral Hutfels, the meticulously constructed *Granite* can be said to be *shoddy*—if stripped of the moral and value judgment now attached to the term—and particularly in the older sense of a garment made out of existing garments, but whose sutures are still visible. By tracing the outline of these spectral stones and scarcely visible seams, by reading the shoddy as the trace of the shudder that marks the *Erschütterung*, Stifter's landscapes acquire a greater vibrancy and texture than has often been acknowledged.

The passages where stone projects into the narrative are sites of menace rather than solace. In *The Pitch Burners* the link between the Hutfels and the vestibule stone is established in a mutual situation of paralysis and petrifaction. This is attested to by the narrator, who describes his situation, upon being punished and deposited on the small vestibule rock by his mother, in words that unmistakably recall the ordeal of Joseph stranded on the Hutfels: "She then let me go, abandoning me, sobbing and crying, on the vestibule stone that stood there for the beating of yarn. . . . This turn of events caused such pain that I could not get up from the stone" (*Dann ließ sie mich los, und ließ mich auf dem Steine des Vorhauses, der zum Klopfen des Garnes dastand, weinen und schluchzend sitzen. . . . Ich konnte vor Schmerz über diese Wendung der Dinge nicht von dem Steine weg*) (*HKG*, 2.1:14). However, with the retouching of *The Pitch Burners* for the publication of *Granite* and the excision of the violent Hutfels episode, the significance of this young narrator being cast out on the vestibule rock would also be lost. And so in *Granite* the narrator instead climbs of his own accord onto the vestibule rock: "Toward this stone I staggered and collapsed upon it" (*HKG*, 2.2:27; *Gr* 10). Notwithstanding these editorial palliatives, both the vestibule rock and the larger boulder before the house retain a menacing character in *Granite*, not only because they continue to play a supporting role in the opening scene of a defilement and subsequent punishment but also because of the way in which the missing stone haunts those that remain.

On this vestibule stone the yarn for domestic weaving is beaten, and the young narrator collapses on it, following his own beating: "In a corner of the entryway there is a large stone cube, and on it the yarn for domestic knitting is beaten with a wooden hammer. Toward this stone I staggered and collapsed upon it. I could not even cry, my heart was crushed, and it was as though my throat were laced shut" (*HKG*, 2.2:27; *Gr* 10).[39] It could be said that the mother, sewing just as he entered the house with his pitch-covered feet, was in the process of adding another pleat to a complex whose Oedipal intimation is more than merely ornamental. The indelible footprints left on the freshly cleaned floors, the trace of the plague they bear—also the backdrop of Sophocles's

Oedipus Rex—the subsequent soiling of his mother's skirt, an absent father, whose existence is mentioned only once in the narrative and even then only in the penultimate paragraph, and the swollen feet of the outcast visible even through the stains as he staggers toward the stone: these are the few significant glimpses of a scene of "turmoil" (*Zerwürfnis*) and "disorder" (*Unordnung*). With respect to this, the narration of the subsequent excursion with the grandfather into the countryside, during which the story-within-the-story of the plague and Joseph's abandonment is narrated, could be considered to be a carefully orchestrated detour, a detour around a soiled domestic space and a silent but seething maternal figure, one whose face remains obscured, whose hands emerge only to thrash the boy and to labor to scrub away the tracks he has left on the flooring, and whose voice is allowed but one malicious utterance in the entire span of the story: "What does this hopeless, obstinate son have on him today?" (*HKG*, 2.2:26; *Gr* 9).

With the Hutfels effectively effaced from *Granite*, a narrative red thread—like the one which the heart of the young narrator in *The Pitch Burners* is drawn closed (*das Herz war mir mit Schnüren zugezogen*) (*HKG*, 2.1:14) and with which in *Granite* his throat is laced shut (*die Kehle wie mit Schnüren zugeschnürt*) (*HKG*, 2.2:27; *Gr* 10)—is thereby undone, no longer linking the needle in the mother's hand with his sutured throat with the hands of the other outcast child, Joseph, frantically unraveling and reknitting his jacket from a boulder that was to be both tomb and tombstone. (Stifter's language also links up this passage one of the most well-known scenes ever penned by an Austrian: Sigmund Freud's dream of Irma in *The Interpretation of Dreams*, where Irma complains that she feels as though her pain has constricted her throat, stomach, and body and has literally sown her shut: "it's choking me" [*es schnürt mich zusammen*], she says.[40]) As for the garment the mother was knitting when her son, the narrator of *Granite*, entered the house with his unseemly feet: it is never resumed in the course of the story. The proliferation of these and other loose ends throughout the text seems to defy the narrative of continuity and containment typically ascribed to Stifter's writing. Isolated from the narrative moment that should have prefigured the present-day ordeal, both the vestibule block and the larger block outside the house assume the position formerly occupied by the boy cast out in *The Pitch Burners*, namely that of an alienated object: abandoned, displaced, shorn of genealogical relations, a foundling (*Findling*).

What the above allusion to Oedipus could bring out here, though, is less an unarticulated incestuous desire, sublimated transgression, or scene of castration—though these cannot be entirely discounted—and more generally the structure of a block whose frustrated articulation can take place only in another scene than that of speech, namely in the lithosphere and those material blocks that occupy the most prominent sites in the story. However, that is not to say that mute, inorganic nature merely stands in as an empty stage on which a human story unfolds. The earth is a medium in which multiple storylines unfold, and one of which involves the menacing agency of the inorganic in a narrative that cannot be contained within a merely human drama.

Ungrounding *Granite*

The spectrality of stone in *Granite* extends to the foundations of the opening block, where it sits uneasily on its sandstone base. Although the deteriorated state of the sandstone does provide evidence for the extended duration of the block in its present location, thereby signaling something of the durability and stability hopefully indicated in the title *Granite*, this can be posited only negatively: "Its age is also proved by the circumstance that the sandstone slabs upon which the stone rests have been completely worn out having been trod upon [*schon ganz ausgetreten*] and where they extend beyond the rain gutter they are marked with deep holes from the dripping water" (*HKG*, 2.2:23; *Gr* 7).

This passage would substantiate the image of granite as everlasting, did it not encounter a block in the inability of the text to offer any positive evidence of the passage of time: in attempting to mark time it can only point to holes. In place of the flagstone foundation (*Unterlage*), which has been ground into nothing, all that remains is a writing that can only reiterate the inexistence of any documentation (*Unterlagen*) of the age of the block.

Furthermore, in the folding together of a ground susceptible to erosion and a writing susceptible to revision, a geological process becomes a poetological principle. Stifter's is less a geopoetics of an impermeable and obdurate granite substratum and much more so of the sedimentary and the erosive. As instances of "gentle law" that frames the preface to *Many-Colored Stones*, Stifter mentions two natural phenomena, "the blowing wind" and "the rippling water," both of which can be identified as the primary forces responsible for erosion (*HKG*, 2.2:10). In this way Stifter's geological imagination offers a glimpse into the truth of realism one that, as Adorno remarks, can only be articulated negatively: "No narrative has ever had a share of truth if it has not looked into the abyss [*Abgrund*] into which language collapses when it tries to become name and image."[41] With regards to Stifter, Adorno's dictum about the abyss can be understood not only with regard to the prominent gorges and ravines in his stories but also a literal *Ab-grund* along the lines of *abduct* or *abject*: as a withdrawal of the ground. The abyss of *Granite*, a story in which there is no granite to speak of, is perhaps most evident in the sparseness of the ground, its porosity, and its susceptibility to erosion. Peter Rosei's observation that Stifter knew or at least sensed that the floorboards of language were too thin and too sparse to furnish an adequate ground can be extended outside of domestic space and into the ground that Stifter similarly sensed to be too vibrant to reliably reproduce the messages and significance he attributed to it.[42]

The relentless dissipation and diminution of things in *Granite* characterizes its erosion imaginary. "'But afterwards other generations came,' he continued, 'who know nothing of the matter and scorn the past; the fences are gone, the places [*Stellen*] are covered with ordinary grass'" (*HKG*, 2.2:45; *Gr* 22). Throughout the story the narrator remarks on a number of spatial changes, principally in the form of reductions: horizons are obscured, over the generations the forests become smaller and smaller as pastures expand, the enclosure of a mass grave by the church has been effaced, the site itself is overgrown, the monument to the plague has inexplicably disappeared, the resting place of the dead is disregarded by the newcomers, and by the end of the story the neighbor's rowan tree "was not half as big as it had been yesterday" (*HKG*, 2.2:60; *Gr* 33), the towering *Mutter* (mother) in the opening pages diminishes in the penultimate paragraph into a wizened *Mütterlein*, and the place names (*Pestweise, Peststeig, Pesthang*), in which the plague (*Pest*) has been inscribed, are spoken with a thoughtfulness that is registered foremost in the ease of their pronunciation: "They don't respect the places where the dead rest and they say the name *Pest* with a thoughtless tongue [*leichtfertiger Zunge*] as though it were any other name" (*HKG*, 2.2:45; *Gr* 22).

There is another sense in which the ground of *Granite* is porous, pervaded as it is by the language and concerns voiced in "A Walk through the Catacombs," a vignette from Stifter's 1844 collection *Vienna and the Viennese*. In an extended preface to the actual descent into the catacombs, Stifter surveys the above-ground displacement of the cemetery in front of St. Stephen's cathedral, a victim of modernization, and the attendant diminishment of respect for the dead: "Their crosses and monuments have disappeared, the praise of their virtues on the same has fallen silent, the memorials that once grounded them, in order to signify the sites of their kin, have been thrust by our industry and traffic up against the walls of the church" (*HKG*, 9.1:51). If the grandfather in *Granite* is hyper-attentive to changes and above all disappearances of certain features in the landscape, if the protagonist of *Indian Summer* seems inordinately concerned with the pile of displaced river rocks (having just set off from Vienna), these responses might be accounted for by the recognition of their being pervaded by the kinds of transformation and displacement that Stifter has witnessed in Vienna. In fact, "A Walk through the Catacombs" opens with an observation of Vienna as a geosocial phenomenon: due to diversified means of social interaction, the inhabitants have become "finer, smoother, more malleable, like pebbles that abrade one another" (*HKG*, 9.1:51). Not only a rhetorical figure, this observation speaks to an affiliation between the human and the mineral, particularly under the sign of their mutual exposure to displacement and abrasion. Stifter is particularly affected by the repavement (with paving stones) of the cemetery next to the church: the preparation for this renovation involved loosening the Earth, which exposed numerous human remains that were then carted away and placed on a heap (*Haufen*). As readers of Stifter have observed, the word fields around aggregates, heaps, strata, and debris (*Häufung, Haufen, aufgeschichtet, Geschiebe*) proliferate through the texts ranging from *Vienna and the Viennese* to *Indian Summer*, where they pertain to anthropological and geological objects alike.[43] And even in the deepest recesses of the catacombs, Stifter re-encounters remains in as much a state of disarray as the surface. At the end of the walk, and in response to the promise of the pastor that the catacombs offer "a stately and undisturbed burial place," Stifter skeptically questions, "as if anywhere on the Earth there was something undisturbed, unephemeral! indeed is the Earth itself not transitory and becomes a corpse like that which one now so carefully shrouds in its stomach?" (*HKG*, 9.1:58).

In "A Walk through the Catacombs" and all of the peripatetic narratives that follow, Stifter overhears the dead occupying land and language and speaking in tongues, all the corpses in these coarse tongues. *Granite* is the most pronounced attempt of any narrative to serve as their tombstone. It would be the fallen memorial, upraised again, but the stone is undermined by the transience that ravages the story and disallows it from repositioning or recollecting what has been disordered and scattered. Sifter's question posed in the catacombs resounds here: *as if anywhere on the Earth there was something undisturbed, unephemeral!* In *Granite* Stifter not only knows or senses but transitively *writes* the inadequacy of language and its inability to obtain the sovereignty over the world that some of his critics see it as possessing. And this falls together with his predilection not for stories that take place on smooth floorboards but rather those that unfold on a ground marked everywhere by its conspicuous vacancies.

The slabs of sandstone below the more enduring block serve less as an index of the relative age of the block and more of this fundamental insecurity of the ground, whether it be lithic or linguistic. That this language has a false floor can be gathered from its status as a reinscription: in *The Pitch Burners* the block rests not on sandstone but on a bed of *tuff*, a porous rock typically formed by the consolidation of volcanic ash.[44] The reason why tuff might no longer be tenable in the revised version published in 1853, and why it would be replaced by a more generic sandstone could be sought both in Stifter's reactionary response to the political revolutions of the intervening years as well as the fertile ground that the uprisings found in the volcanic imagination of the nineteenth century. By the 1840s the simplistic, ideologically driven geological debates between so-called Neptunists and Vulcanists were being challenged by far more complex and dynamic geomorphological models, including Lyell's uniformitarianism, yet theories of catastrophism based on the prominence of volcanic action in the construction of the Earth still found an increasing number of adherents both in academic circles and the European revolutionary imagination, especially around 1848. Carl Vogt's claim, in his 1847 inaugural lecture as Chair of Zoology at the University of Gießen, that geology is nothing but "the revolutionary history of the planet" (*die Revolutionsgeschichte des Erdballs*) has an important pretext not only in his support of radical social reform but also in Alexander von Humboldt's 1823 lecture, "Concerning the Structure and Action of Volcanoes in Different Regions of the Earth" ("Über den Bau und die Wirkungsart der Vulkane in den verschiedenen Erdstrichen"), the work that was a source of such consternation for Goethe, and which, by virtue of being included in *Views of Nature* (*Ansichten der Natur*), was enjoying a wide readership by its third edition in 1849.[45] The increasingly significant role played by volcanic and tectonic activity in geoscientific theories of the Earth substantiated in particular those calls for social revolution that came from figures, like Vogt and Bakunin, who postulated the identity of nature and society in the act of revolution. And so, placing the block on a pedestal of volcanic origin, in a prevailing political allegorese, would have undermined the attempt to establish continuity "beyond the catastrophic, stability beyond the destructive moment." It would signify quite the opposite.

For a Lituraterre

As much as the block at the outset of the story is said to enshrine an irrevocable hegemony and provide the emblem of an enduring status quo, the writing notably fails to carry out this act of enshrining. If the block at the outset of *Granite* is taken as an emblem of anything, it is not one of material permanence and the obdurate persistence of the status quo but rather one of erosion and abandonment that is intertwined with Stifter's process of writing. The widely accepted reading that "the block of granite stands as the emblem of 'rock-solidity' enshrining the continuity of both the natural and the social realm," itself encounters a block, first in the absence of any identification of the substance of the stone and then in the resistance of this stone to any form of commemoration or inscription.[46] On the contrary the block does not stand for anything—in the first place it *lies*: the text emphasizes its horizontality—if not the inability to tease apart the geologic and the social. For one, in not standing, and in not being able to stand for anything, the block stands for an ambivalence that stems from but cannot be reduced to Stifter's response to the March 1848 uprisings, an ambivalence depicted in the nonconformities and inconsistencies resulting from the heavy revision and retouching of the narrative over these years.

Where disorder emerges in the story, first in the fouling by the boy's footsteps of the floor and secondly by the historical narrative of the plague, it does so without any overt connection to the widespread social unrest of the time. But in the grandfather's apparent inability to point to any enduring physical memorial of the social upheaval that came in the form of the plague, the text reflects its own speechlessness with respect to the 1848 uprisings. This speechlessness is also gently addressed by Stifter in the euphemistic "experiences of the recent year" (*die Erlebnisse der letztvergangenen Jahre*) mentioned in the preface to *Many-Colored Stones* (HKG, 2.2:17). Although readers have registered this historical caesura in the revisions made between the stories collected in *Many-Colored Stones* and the previous journal editions, they have tended to do so at the level of the signified, referring to elements such as the expansion of idyllic storylines and the omission of more tragic and violent events. Yet traces of the "experiences of the recent year"—no matter how inexplicit, indirect, and illegible—also materialize in *Granite* in the vacancies and gaps in the landscape and narrative alike. The same could be said of his recent experiences, addressed more explicitly in *Vienna and the Viennese*, of environmental upheaval through urbanization. Those telling gaps and inconsistencies in *Granite* are precisely the interstitial space of shock that Walter Benjamin identifies when—returning to the matter of "shock" (*Erschütterung*) several decades after his essays on Stifter—he resolves it into the word for "thin" and "sparse" (*schütter*).[47] There is a latent shock of the Anthropocene in Stifter, traces of which can be gleaned from the discarded pages that, although not taken up into the canonized versions of his stories, persist like the titular block in the form of superfluous objects and inconsequential storylines.[48] Their persistence disrupts Stifter's attempts to neutralize "disruptive nature" in his stories by means of what Sean Ireton identifies as a robust conceptual and narrative pattern of "order—disruption—reinstated order."[49] In resisting permanent inscription and in resisting inscription into any sequential chronology or stratigraphy, the block resists the conflation of human history with processes of layering and consolidation. To open the question of history in *Granite* is to open—to suspend—any significance to the etymological relationship between strata (*Schichte*) and history (*Geschichte*).⁵⁰ The history of the text of *Granite*, for one, is less one of consolidation and more one of a loose agglomeration, constantly susceptible to settling, or *unsettling*, in which the discarded scraps of writing do not dissolve into the final version but instead haunt the published versions in the spectral form of unconformities and inconsistencies. Reading the discarded passages not as stations in a teleological trajectory but as stand-alone text blocks casting shadows on later versions is supported by the broken teleologies showcased in Stifter's stories.⁵¹ Moreover, these unconformities provide a model for thinking of history outside of the linear stratigraphic process of a conventional historicism and in doing so contribute to recent critiques of the stratigraphic logic of historiography.⁵² What this close reading of Stifter hopes to have contributed to that conversation is a model and method of decrystallizing history that would counter the assertion of a false totality in geologic structures of consolidation, stratification, and sedimentation. If literary structures are read as the precipitate of a historical consciousness, it might be precisely in the faulting and rupturing of stratigraphic readings—in the seams—that this consciousness is active in *Granite*.

A closer look at the conspicuous absences around the block makes evident the resistance of this stone to any form of commemoration or inscription. *The Pitch Burners*, more so than *Granite*, attempts to inscribe the stone within a stabile social system: "If each of our predecessors had inscribed a distinguishing marker [*Merkmal*] on it, the stone would have been be a memorial marker [*Denkzeichen*] of our forefathers reaching far back in time" (*HKG*, 2.1:11). The inability of this passage to posit anything more than an irreal inscription ("the stone would have been") attests to an alterity and instability that overwrites the material qualities of granite with a radical mutability of literary writing that is evident in the two draft variations:

> If each of our predecessors had inscribed a distinguishing marker [*Merkzeichen*] on it, the stone would have been be a monument [*Denkmal*] of our ancestors reaching far back in time. (*HKG*, 2.3, 125)

> If each of our predecessors had inscribed a distinguishing marker [*Merkzeichen*] on it, the stone would have been be a memorial marker [*Denkzeichen*] of our ancestors reaching far back into prehistory. (*HKG*, 2.3, 125)

In exhausting the combinatory logic of "*Merkzeichen*" and "*Denkmal*" (*Merkzeichen—Denkmal—Merkmal—Denkzeichen*) the story is one of the inability of any designation to stick, any monument to be erected, any instrumentalization of the stone for commemoration. (This will be even more evident in the published version of *Granite*, which goes a step further by avoiding any of these irreal invocation to the memorial.) What remains, in the absence of any inscription except for that of its resistance to any script or scripting, is an unremarkable stone whose inappropriateness as an "image of an enduring site in space" is evident in this proliferation of passages around this prominent impasse.[53]

The finding that the stone "preserves, since time immemorial, the traces of generations" cannot be attested to by *The Pitch Burners* or *Granite*.[54] The block no longer appears to function as a reliable recording medium or other source with which to store or otherwise restitute lost time; it has been dubbed over too often. The only trace of the age of the stone is its utter lack of texture, and this remarkably polished stone offers an emblem for a philological practice of glossing over inconsistencies in his stories. In the passage describing the stone not being inscribed, granite cannot stand as an emblem of human duration. What persists in time is neither the inscription nor the obdurate stone but rather its twofold resistance to being inscribed *upon* and *in* writing. The stone does not endure in space as a *stele*, an upright slab covered in inscriptions and commemorating an event or figure in their collective memory. Such an object could have been found in the columns erected in the town square of Oberplan to commemorate the plague, and on which one could have read when the plague came and when it ceased, had that memorial not disappeared as well.[55] In place of a stele is a textureless surface that furnishes, more than a literal setting for the story, a screen for literature to project its imagination of itself as a *lituraterre*, Jacques Lacan's marker of the complex relationship between literature and terrestrial systems (*la terre*) in a time of mutual obliteration (*litura*).[56] With the publication of *Granite* even the description of the impossibility of inscription has been effaced. This is all the evidence that can be mustered, over numerous rewrites, to indicate the age of the stone: "its upper surface has become so fine and smooth from much sitting, as though it had been covered with the most artful glaze" (*HKG*, 2.2:23; *Gr*, 7). Even the description of the erasure of any texture is posed in the subjunctive mood. In a word, the stone is a palimpsest—that which has been repeatedly rubbed smooth—of traces of obliteration.

That stone is obstinately illegible and resistant to permanent inscription need not contradict the writing that proliferates around it, nor does it place in question the decision to base the entire narrative around it. As Eva Geulen suggests, Stifter's physical sites, his *Stellen*, are not places of refuge for an extralinguistic substantiality otherwise lacking in the oral story; on the contrary, "they are much more abysses, in which the language of the texts opens out toward what is unavailable for it."[57] What is unavailable for *Granite*? An immutable language, society, and earth. In this way *Granite* offers an image of the Earth contiguous with the increasingly dynamic image rendered by the contemporaneous earth sciences and social movements alike. But as Tove Holmes observes, the unavailability of permanent inscription, just like the unavailability of solid ground, does not preclude archival practices from taking shape. "An archive of the Earth as it continually registers the effects of culture while providing its shifting epistemological substrate," writes Holmes, drawing on Foucault's concept of the archive "would necessarily be similarly incomplete and open-ended."[58] As a story that finds itself unable to speak of contemporary political events, *Granite* lights upon an adequate emblem in the block of granite, albeit not as an emblem of the continuity of history but rather of a history of displacement, decipherment, erosion, and transformation. Rather than the story appropriating the apparent obdurateness of granite and conveying this emblem of persistence onto the decidedly more fluid familial and societal institutions, the text finds within stone a recapitulation of its own impermanence and formal incompleteness.

4
The Shock of the Earth
Benjamin's Unarticulated Ground

Scattered throughout the writings of Walter Benjamin is a figure of shock distinct from the one that characterizes the visual impressions and electrical discharges peculiar to the experience of moving through the modern city. Sparse remarks dispersed across several decades stage a shift from theorizing such sudden disturbances toward registering a deeper time of collapse. Not the city per se but the Earth becomes the locus of a shocking experience that can be placed over and against what Benjamin calls *Chock*, which is both urban and urbane, which operates at the human scale of perception, which is both sudden and anticipatable, and which, in the image Benjamin lifts from Baudelaire, the man of the crowd parries away like the blows confronted by a graceful fencer. Benjamin's walks within and without the city register a different kind of tremor. Like the little hunchback haunting *Berlin Childhood*, the multitude of shocks, shake-ups, tremors, concussions and shattering things described by the word *Erschütterung* are insidious, subterranean, imperceptible yet cataclysmic, and accessible only retrospectively. Typically translated in the *Selected Writings* as "shock," *Erschütterung* discloses a vibrancy of language that surpasses any given word. Its range extends from deep in the lithosphere and the tectonic "tremor" to a somatic "shudder" in its mythological, theological, or psychological manifestations (primordial "shuddering," mystical "trembling," and existential "shock") to aesthetic experience ("being moved" by art) and out into the atmosphere and the sonic "concussion." Although this "shattering" is a verbal event for Benjamin—the shattering of the word—its material relays extend across multiple forms, media, and scales. The *Erschütterung* marks a derangement that it is especially noteworthy in an epoch characterized, for its ecological shortsightedness, as the Great

Derangement.[1] Yet in readings of Benjamin this figure has been all but eclipsed by a form of shock whose aesthetics of suddenness has been appropriated in the service of a "narrative preference for a retrospectively sudden or instantaneous trauma with condensed duration—the comet impact, the abrupt fault, the nuclear blast, the successful revolution, and so on," as Tom Cohen notes. "This imaginary, reinforced by phenomenological bias ... occludes metrics other than hypertechnic speed, the cataclysm that arrives in cinematic slow-motion over decades or more, concealing numerous feedback loops and invisible extensions."[2]

Now might be the time to unearth this isotope of shock. In terms of this chapter, the temporality, geomorphism, and geoaffectivity of the *Erschütterung* offer an aesthetics of and for the mineral imaginary. Moreover, in terms of a contemporary epoch marked by transgenerational trauma and other forms of slow violence—Rob Nixon's term for political emergencies of the long-term such as those cascade effects triggered by deforestation, the use of radioactive weapons, and climate change—this allo-chronic and allo-anthropic mode of shock promises to better address the representational and conceptual challenges posed by the incremental, often anonymous, and imperceptible threats endangering life in the Anthropocene.[3]

The unavailability of the language in which geologic processes could be registered and gathered into a recognizable event has been acutely perceived by the nineteenth-century writers who have been the focus of the study: the excessiveness of the Earth in Tieck and other romantics; the erraticism of the lithosphere in Goethe; the unreliable ground of Stifter's many spectral stones. Closer to the present day, but not far removed from the nineteenth century, Walter Benjamin's elaborations of shock, which evince an astonishing sensitivity both to the unsettled earth underfoot and planetary irregularities alike, offer a way to figure the disorientation of Tieck, Goethe, and Stifter—as well as that of a contemporary epoch facing planetary collapse. "While the global environmental crisis intensifies," Timothy Clark recently observed, "the inherited language with which it is often addressed has itself become more fragile."[4] Today the keystones of the environmental imagination—including nature, the Earth, the pastoral—no longer possess an intact physical reference. At the same time, the hallmarks of modernity, and this chapter singles out the distinctive aesthetics of suddenness and shock, are inadequate to account for experience in the Anthropocene, which Bruno Latour glosses as "the most decisive philosophical, religious, anthropological and . . . political concept yet produced as an alternative to the very notions of 'Modern' and 'modernity.'"[5] Relatedly, in the above-cited passage, Tom Cohen places the contemporary world in a state of aftershock: after the metaphorics of "shock" that would allow for the identification and moreover the containment of a discrete event. Although Christophe Bonneuil and Jean-Baptiste Fressoz's *The Shock of the Anthropocene* (*L'événement anthropocène*) seems to signal otherwise already in its title, their argument works against delimiting the Anthropocene to a definite, measurable event; identifying it as such for them only means "deconstructing" official accounts and "forging new narratives . . . and thus new imaginaries. Rethinking the past to open up the future."[6] Along similar lines, rather than outright discarding shock's capacity to account for a planetary crisis, this chapter proposes to salvage and articulate a variant that was not entirely biased toward the scale and rhythms of the human body. The crises in representation to which Nixon, Clark, Cohen, Bonneuil, Fressoz, and others draw our attention furnishes a now of recognizability that brings into focus Benjamin's elaboration of an alternative form of shock; Benjamin himself was writing from a present whose immense scale of destruction he characterized as a global catastrophe, or *Weltuntergang*.

The Unconsolidated Ground of Shock

From the resonant asphalt of Berlin's Tiergarten to the porous rock of Naples to the uneven paving stones of Paris and the unarticulated ground of Marseilles's public squares to the rocky paths of Ibiza, the ground from which Benjamin's thought proceeds is an unarticulated one: irregular, uneven, rough, provisional, fractured. That no pronounced theory of the *Erschütterung* emerges in Benjamin's writings is related to this unconsolidated ground. The seismological associations of the word also contribute to its ongoing instability. Up through the eighteenth century an earthquake was commonly known by the more descriptive *Erderschütterung* (both a shaking motion and a shattering action) rather than the tectonically accurate *Erdbeben* (strictly a quaking motion). By the twentieth century it had come to account for subjective experience in particular, where it did not become the cornerstone of an aesthetic theory until, posthumously, Theodor Adorno elaborates a phenomenology of aesthetic receptivity in his *Aesthetic Theory* that is largely derived from Benjamin's scattered writings on the subject. In a key passage, Adorno writes: "The shock [*Betroffenheit*] aroused by important artworks is not employed to trigger personal, otherwise repressed motions. Rather, this shock is the moment in which recipients forget themselves and disappear into the work; it is the moment of being shaken [*Erschütterung*]. The recipients lose their footing; the possibility of truth, embodied in the aesthetic image, becomes tangible [*leibhaft*]."[7] The absence of a discrete aesthetic theory of the *Erschütterung* in Benjamin's writings can be understood other than as a deficiency to be later redeemed, even when Adorno's description largely accords with Benjamin, except in one key aspect. Those places were the *Erschütterung* outcrops, from the early writings on the ethics and philosophy of language to the thought images (*Denkbilder*) of the late twenties and literary criticism in the later thirties, suggest a prolonged engagement with a protracted form of shock not necessarily accessible through the museum, gallery, the book, or any visual media. For Benjamin the aesthetic experience signaled by the moment of being shaken is facilitated by a literal loss of footing and is in most cases not triggered by an artwork.

Already the earliest notes on the *Erschütterung* are tied to a loss of articulation vis-à-vis the physical world. These writings emerge out of Benjamin's extended engagement, beginning in late 1917, with the nineteenth-century Austrian writer and pedagogue Adalbert Stifter, whose literary depictions of a natural world subjected to a "gentle law" (*das sanfte Gesetz*) often belies the catastrophism of that world.[8] The most developed writings, the sketches known as *Stifter I* and *Stifter II*, take the form of a moral and an aesthetic argument against Stifter, respectively. The argument in *Stifter I* might be summed up as Stifter is too earthly. In Benjamin's thought in this period, the highest realm of justice is an unearthly realm dematerialized and purified of any relation to the profane and the everyday. It is a realm of fate as opposed to character, justice as opposed to life, the great as opposed to the small. The extensive descriptions of natural phenomena in Stifter's stories, Benjamin argues, are secularized and do not furnish an eschatological dimension of "final things" necessary for justice. By linking fate with the vicissitudes of the natural world, Stifter is unable to present the large-scale relations which concern metaphysics.

The aesthetic argument of *Stifter II* builds on that of *Stifter I*. Stifter, writes Benjamin, is so immersed in the physical world that he has no capacity for perceiving or presenting revelation (*Offenbarung*), since this belongs to the metaphysical-acoustic sphere of the word. Consequently, stillness (*Ruhe*) is a fundamental feature (*Grundzug*) of Stifter's writing, which only engages the visual sphere. Therefore, Benjamin argues, Stifter's spiritual muteness results from an inability to present shock, as it has an irrevocably acoustic dimension. It is on these grounds that his argument begins to unravel, though tellingly. Whether or not Benjamin's criticism is valid is perhaps less interesting than seeing how he himself struggles to articulate this form of shock—and thereby articulates what will becomes its fundamental feature: "The ability to somehow present any deeper emotion or 'shock,' whose expression man primarily seeks in language is absolutely lacking in him."[9]

Even in Benjamin's speculation the question of how to present a shattering experience has no straightforward answer. Already his expression "somehow or another" indicates a blockage in the attempt to articulate a specific technique for its presentation in language, and by further placing "shock" in scare quotes, he exposes a certain uneasiness that extends beyond his powers of articulation. This frustration of expression might best express the *Erschütterung*, taken as a figure of unsettling movement that is itself highly unsettled. For it is not "shock" itself that eludes Stifter, or Benjamin for that matter, but instead the ability to place it in language. The word *Erschütterung*, rather than accounting for an existential "shake-up" or discrete seismic event, would seem to find its presentation only in what never ceases shuddering, shattering, and stuttering, what remains unarticulated, whether that is language or the Earth.

As revealed in chapter 3, the *Erschütterung* of and in Stifter's texts can be sought in the debris (*Schutt*) that marks his geologically active landscapes. Increasingly for Benjamin articulating shock becomes tied to the problem of an unconsolidated ground. "It is worth pointing out," he writes with respect to the poem "Against Temptation" ("*Gegen Verführung*") in his commentary on poems of Bertolt Brecht, "that the word *Erschütterung* ['shock'] contains the word *schütter* ['thin,' 'sparse']. Wherever something collapses, rifts and gaps appear. As analysis has shown, the poem contains numerous passages in which words combine in a loose, unstable way to form the meaning [*zum Sinn zusammentreten*]. This contributes to its shocking [*erschütternd*] effect" (SW, 4:224; GS, 2.2:548–549).

Schütter, the word inside the word for shattering, hardly signifies an obdurate etymological kernel that would signal a terminus in which meaning could come to a rest or otherwise cease proliferating. Meaning "thin" or "sparse," a material defined by interstitial openings rather than its substance, the resolution of *Erschütterung* into *schütter* facilitates rather than prevents the unsettling of meaning associated with the tremor. The word recalls what in set theory is known as *nowhere dense*: a topological space with an empty interior. The more you zoom in, the more anything substantial or differentiable recedes. In other words, those spoken by Benjamin's slow, deliberate voice (citing a line by Karl Krauss) to Jean Selz while they are sitting on the terrace of a waterside café in the port of Ibiza: "The more closely you look at a word, the more distantly it looks back" (*Je näher man ein Wort ansieht, desto ferner sieht es zurück*).[10] The moment of the *Erschütterung* moves through language in the form of a word-shattering earth and an earth-shattering word. From a distant vantage point they become indistinguishable: e r sh tter ng.

The shocking effect of this Brecht poem extends into Benjamin's writing. He does not just write that the words assemble in a loose, unstable way to make sense (another meaning of *Sinn*) but rather that they stride together toward meaning (*zum Sinn zusammentreten*), and moreover in a way that directs attention away from any particular meaning and instead toward a volatile signifying surface. The volatility of this assembly is attested to elsewhere by Benjamin's incendiary definition of an image as "that, wherein what has been comes together in a flash with the now to form a constellation" (AP, 462; GS, 6.2:578). Similarly, in the movement of meaning in the phrase *zum Sinn zusammentreten*, the words *zum Sinn* ("toward meaning") seemingly break down and reassemble into the word *zusammen* ("together"). Labile and unstable, words assemble to form a meaning that emerges only in its continual deformation, and their unsteady stride both evokes and revokes the original mathematical definition of *Sinn* as "direction." For the action of assembling (*zusammentreten*) can also be a violent one; it can entail *treading* on someone or something repeatedly so as to provoke the collapse (*zusammenstürzen*) of all its parts at once. Although this activity could be said to characterize the critic's engagement with Brecht's language, applying pressure to a word until it breaks, in Benjamin's account, words themselves are apparently the agents of this disassembling of meaning.

Language, and this is a conclusion in the Brecht commentary, is the scene of a conflict whose rifts and gaps commentary can expose but not mend. Not only in this regard is commentary closely related to translation. In the final paragraph of "The Task of the Translator," Benjamin observes a similar instability in the late Hölderlin: "The Sophocles translations were Hölderlin's last work. In them meaning [*Sinn*] plunges from abyss to abyss [*von Abgrund zu Abgrund*], until it threatens to become lost in the bottomless depths of language [*in bodenlosen Sprachtiefen*]" (SW, 1:262; GS, 4.1:21). The revelation not only that language could be without ground but that it could actively *unground* is contained in this account of Hölderin. But it can even be observed in the English translation of the above passage from the Brecht commentary in *Selected Writings*, particularly the line "the word *Erschütterung* ['shock'] contains the word *schütter* ['thin,' 'sparse']."

What the *Selected Writings* does not carry over from German is a stable meaning resulting from a harmonious assembly of words. It is telling that the word that follows "the word" in the translation is not a word, at least not one that exists in the English language. Twice in the course of a single sentence the translation breaks down: in one phrase ("the word *Erschütterung* ['shock']") and then again in the next ("the word *schütter* ['thin,' 'sparse']") a chiastic reversal occurs, in which the German words enter into the English text, whereas the English words are set off in brackets and quotation marks. The translation interiorizes something unintelligible and radically exterior to it, exemplifying a traumatic shock. Yet for Benjamin the act of translation would be maintained precisely in these moments where it carries over those breaches that emerged where language shuddered and the word shattered. Even where translation glosses "*Erschütterung* ['shock']" it does not gloss it over.

The Geo-aesthetics of Getting Stoned

If the experience of the *Erschütterung* is not one that can be so easily articulated, then, as is the case in Stifter's peripatetic narratives, the shock of and in Benjamin's writing could be sought in what remains unarticulated in it: in the ground underfoot that never fails to perturb his writing. Even though the physical ground most frequently appears in city-dweller Benjamin's writings as something paved or otherwise surfaced, it is anything but impermeable and unshakeable. It is resonant and resonates with his footsteps: in a word, it is vibrant. These vibrations facilitate deviations from those lines of inquiry not attending to the physical ground, and in doing so they offer a perhaps unlikely source of inspiration for a geophilosophy critical of the ways in which, as Ben Woodward writes, "the Earth has been used to ground thought instead of bending it."[11] Mineral matter is the object of Benjamin's studies ranging from the baroque emblem studies in *The Origin of German Tragedy* to the architectural investigations of *The Arcades Project*, but it is particularly in the writings of the late twenties and early thirties and the period encompassed by the unfinished *Arcades Project* that the capacity of stone and earth to unground lends them a prominent place in his methodological reckonings. As he writes in a 1928 letter to Gershom Scholem regarding the surrealists, a remark which will later be recapitulated in *The Arcades Project*: "I am gradually starting to come more and more frequently upon passages of young French writers, who, while pursuing their own lines of thought, their own trains of thought, betray only fluctuations, aberrations [*Schwankungen, Irrungen*], yet the influence of a magnetic north pole that disturbs their compass [*ihren Kompaß beunruhigt*]. And I am steering straight for it."[12] The point of reference for these considerations encompasses both French surrealism and the realism of Adalbert Stifter, whose preface to *Many-Colored Stones* famously calls attention to the local fluctuations of a magnetic needle against a planetary backdrop in which these changing values are revealed as coherent waves. Exposing his thought to those wandering electromagnetic fields generated by the movement of molten iron alloys in the Earth's core, Benjamin here acknowledges the planetary scope of his writing and the importance of the aberrant for his thought.

One way of accessing this perspective involves getting stoned. "The hashish eater's demands on time and space . . . are regal," writes Benjamin in the protocol of his hash experiment on September 29, 1928, as they offer access to "immense dimensions of inner experience, of absolute duration and immeasurable space" (SW, 2.2:674; GS, 4.1:410). In these drug experiments his attention is directed to the stones and earth underfoot, but they swell into the widest expanses; likewise Benjamin crosses over the entrenched binaries of the city and the country, the flâneur and the wanderer: "On the way to the Vieux Port, I already had this wonderful lightness and sureness of step that transformed the stony, irregular earth [*steinigen, unregulierten Erdboden*] of the great square that I was crossing into the surface of a country road along which I strode at night like an energetic hiker" (GS, 4.1:411).[13]

Scenes of crossing-over will be familiar to readers of the *Arcades Project* in the glass and steel arches connecting opposing buildings, the arcades' entanglements of interior and exterior, local and foreign, but what prompts the leap here is a "stony, unregulated earth" (*unregulierten Erdboden*) or what, in a revision published in December 1932 in the *Frankfurter Zeitung*, he calls a "stony, unarticulated earth" (*steinigen, unartikulierten Erdboden*). In these turns of phrase Benjamin does not engage in a mystification of the Earth; instead, he calls attention to the material inconsistencies of the ground—cryptic stoniness, irregular roughness, and aggregative unevenness—and their resonances, as indicated in a footstep's ability to trigger the cultural memory of a romantic wanderer, not to mention those (im)personal *mémoires involuntaire* which the act of tripping over uneven paving stones activates for Benjamin no less than Proust.

Benjamin is struck by the resonance of the irregular, unarticulated ground, and as the effects of hash start to hit him, the street transforms from a place of passage and into a gaping wound—"The street I have so often seen is like a knife cut" (SW, 2.2:674; GS, 4.1:410)—that threatens to engulf him: "I immersed myself in contemplation of the sidewalk [*das Pflaster*] before me, which, through a kind of unguent [*Salbe*] with which I glided over it, could have been—precisely [*als eben dieses Selbe*] as these very stones—also the sidewalk of Paris [*auch das Pariser Pflaster*]" (SW, 2.2:677; GS, 4.1:414–415).

The appearance of the street as a wound establishes it as the site of a cryptic trauma, and the play of language in this passage exposes how this ground has been encrypted. Benjamin does not write that he immerses himself in contemplation of a sidewalk; instead, he writes that he immerses himself directly *in* those paving stones known in German as *Pflaster*. This is a form of surfacing with more dimensions and resonances than those furnished by the portmanteau "sidewalk." Making the resonances of these paving stones audible and inquiring into how they furnish a space for his immersive contemplation entails tracing the contours of an underlying topography more convoluted than any simple walkway or any single language could afford. The drifting of the text from sidewalk to unguent retraces the derivation (and deviation) of *Pflaster* from the term for a dressing applied to a cut or wound, a sense still audible in the British English use of "plaster" for bandage. The street could appear as a knife wound because its surfacing is already conceived of as the dressing of a wound, a resonance that would not have been lost on Tieck and the shuddering earth in *Rune Mountain*. In walking through Marseilles, Benjamin traces this fluid movement of the pavement, from stone to liniment, in a way that at once suspends any intact reference to a physical ground but also to an ultimate etymological ground of language. This gliding movement, this series of displacements, continues past the historical derivations of the word *Pflaster* (from *Pflaster* to *Salbe*) and on to a purely phonetic play (from *Salbe* to *Selbe*) before vaulting to Paris and the *Pariser Pflaster* (translated as "the paving stones of Paris" but also containing an echo of the French *plâtre de Paris* or "plaster of Paris" [German: *Gips*], that is, both a type of pavement and a type of dehydrated gypsum historically used for setting broken bones). This movement is facilitated by the irregularities of the physical ground underfoot no less than the volatile signifying surface of language. Aided by this mutual susceptibility to shattering, and beset by an extraordinary case of the munchies, Benjamin's local observations ("Hashish in Marseilles") open out to a planetary scope: "One often speaks of stones instead of bread. These stones were the bread of my imagination, which was suddenly seized by a ravenous hunger to taste what is the same in all places and countries" (SW, 2.2:677; GS, 4.1:414–415).

And yet, the suspension of any intact reference to a fixed physical ground itself reproduces the material property of the ground as something *unarticulated*, in the sense of something spatially undifferentiated, provisional, and liable to change. To say that Benjamin's writing proceeds from an unarticulated ground means that attention must be paid to the writings on physical inconsistencies in a ground not yet nor no longer consolidated into a uniform surface, and it means that his thought is drawn to those physical inconsistencies as a trace of what has not consolidated into an unshakeable word, concept, figure, or form.

Aberrant Thought Images

Among those passages in which words and bodies aberrantly traverse an irregular ground one text stands out: *"Bergab"* ("Downhill"), a miniature prose piece written on Ibiza in 1932, the year in which he reworked and published "Hashish in Marseilles." "Downhill" arguably contains Benjamin's most significant reflection on a geomorphic form of shock. Its location is not unimportant. Neither Paris nor Berlin nor any other twentieth-century metropolis provides the setting. Instead, it is a small island off the coast of Spain, then far off the tourist track, its first hotel still under construction when Benjamin arrived in April 1932. It does not go without saying that Benjamin's thinking and essayistic writing on this island accords with "archipelagic thought" in the sense that Martinican writer Édouard Glissant has elaborated: "With continental thought the mind sprints with audacity, but that fact makes us think that we see the world as a bloc, taken wholesale, all-at-once, as a sort of imposing synthesis, just as we can see, through the window of an airplane, the configurations of landscapes or mountainous surfaces. With archipelagic thought, we know the rivers' rocks, without a doubt even the smallest ones."[14]

The long walks Benjamin will take on the unfrequented and overgrown paths on this sparsely inhabited island give shape to a peripatetic, archipelagic thought in the Glissantian sense: essayistic, erratic, discontinuous, opposed to systems thinking. "For a while now, with all of my books and writing, I have been sedentary," Benjamin writes in a letter to Gretel Adorno in the period during the writing of "Downhill" and other pieces in the "Ibizan Suite" collection, "and just in the last days I have emancipated myself from my strip of beach and undertaken several large, solitary marches into the vaster, more solitary region."[15] The subtle militarization of these walks as emancipatory marches might be accounted for by his reading material of the time, as he reports in the same letter having just completed the first volume of Trotsky's *History of the Russian Revolution*. But the emancipation that this letter most immediately concerns is an emancipation from the *manus*, the hand, the handwork of reading and writing, and the attendant sedentary lifestyle. Instead Benjamin writes of the *footwork* of the long walk. These walks surpass any other form of intellectual labor in developing a consciousness of time and place. Of them he writes: "Only then do I become clearly aware of being in Spain."[16] Benjamin thinks not only on his feet but also through them, not only on the Balearic archipelago but through it.

The period of his visit to Ibiza, lasting just under three months, was a prolific one. That he will reside on the island in the company of an old friend from university in Munich—Felix Noeggerath, whom he refers to in earlier letters as a universal genius and who counted among his teachers in Berlin Georg Simmel and Ernst Cassirer and in Marburg the neo-Kantian Hermann Cohen—will no doubt contribute to the breadth of his interests on the island.

In the period between his arrival in late April and his departure at midnight on July 17 on a ship to Marseille that had already pulled up the gangplank when he arrived for boarding, he maintains a steady correspondence with Gershom Scholem. He writes a number of literary sketches collected under the heading of *Spanien 1932*, several of which, including *Ibizenkische Folge (The Ibizan Suite)*, would be published that summer in *Die literarische Welt*, the literary supplement to the *Frankfurter Zeitung*. He also begins in earnest to make literary sketches of his childhood memories in *Berlin Chronicle*, and in doing so returns to the miniature prose form, known as the "thought image" (*Denkbild*) that characterizes his writings in *One-Way Street*, and writes the first draft of what will become *Berlin Childhood around 1900* (written largely on Ibiza on a separate trip in the following year). For the first time he brings together, under the heading "Selbstbildnisse eines Träumenden" ("Self-Portrait of One Dreaming"), records of his dreams that he has been keeping and occasionally publishing since the late twenties, including at least one text written while on Ibiza, "The Chronicler." He also reports picking up Proust again after the several-year hiatus that ended with "On the Image of Proust," a reading experience that prompts the writing of "Downhill" as well as a brief talk to be delivered on his fortieth birthday, July 15, 1932, which is a particularly ominous day, since he seems to allude in a letter of June 25, 1932, to Gershom Scholem of plans to commit suicide on this date in Nice. Instead, still in Ibiza, he writes "In the Sun" ("In der Sonne"), a longer prose piece that contains in the penultimate paragraph one of the most pronounced and oft-cited conceptions of his Messianism.

This is just as much a period of new beginnings as it is of provisional endings. His initial observations of the island are characterized by a similar vacillation. Ibiza's economy circa 1932, with its preindustrial forms of agriculture, is entirely archaic, Benjamin writes, but each observation is marked by a "still [*noch*]" that betrays an impending curtailment of those practices: fifty years ago the island *still* did not know of bread; today there are *still* only five or six cows, the island is *still* irrigated according to the an old Arab method of bucket wheels drawn by mules; grain is *still* threshed by the hooves of horses, and the paths are *still* unfrequented (GS, 5.1:448). But in all of these *noch* one hears the approach of a *nicht mehr (not anymore)*. Just as the gesture of the *Berlin Childhood* that Benjamin will begin to write from the preemptive exile afforded by the island is that of a hand waving goodbye to all that, so too is the island itself the site of an immanent rupture of tradition. "But already," he continues, the first "unfinished hotel buildings are standing in Ibiza and San Antonio, which hold out the prospect of running water for foreigners"; the "time to their completion," he adds with a twist of irony, "has become precious" (GS, 5.1:448). One gets the impression that time, unlike the newly installed tap water, is running out.

Running counter to these impressions is Benjamin's conception of time as something running out in all directions at once. Although the suite of writings composed on the island—from *Spain 1932* to "The Chronicler" to *Berlin Chronicle* to the final diary entry detailing Benjamin scrambling over the railing of his ship as it departs—deal with a crisis of time, none registers these crises, these multiple turnings of time, more pointedly than "Downhill." Like the others of the time, it is a text written from the insular space of a diminishing present, in which the entirety of Europe was crystallizing into a totalitarian nightmare. That the setting of the second half of "Downhill" is a mountain is not insignificant in this regard. Taking place off the grid—on an archipelago, outside of an urban setting, outside of the rigid framework of space reproduced in the ornamental grills that populate the prose's writings ranging from *One-Way Street* to *Berlin Childhood* and that act as figurations of the layout of the page—this place offers an appropriate setting to present the turnings and returnings of time.

Despite the remoteness of the Spanish island from the German metropolis, and despite the remoteness of the archaic island from modernity, this solitary walker will recast himself and the mountainous landscape with an explosive energy that can be understood in relation to and as a training ground and rehearsal space for the revolutionary body Benjamin sees developing in the urban centers of the interwar period. His retreat therefore can be read as a form of resistance: his absence from Germany, he writes, is the only reasonably way to honor the opening ceremonies of the Third Reich. Similarly, the emancipation of his writing from traditional academic forms refashions the experimental *Denkbild* form as another primary site of resistance to the threat of an intellectual and even organic disintegration. As he writes to Scholem just after leaving Ibiza on July 26, 1932: "The literary forms of expression that my thought has forged for itself over the last decade have been utterly conditioned by the preventive measures and antidotes with which I had to counter the disintegration constantly threatening my thought as a result of such contingencies" (*Correspondence*, 396; GS, 5.2:1096).

These literary forms of expression are the prose miniatures collected under the *Denkbild* title as well as the myriad small forms, such as notes and diary entries, of which the period on Ibiza is exemplary. These small and short pieces can be understood as a writerly strategy dealing not only with a precarious present but also a future that does not promise to conform to any existing trajectory, *against* which his writing would then principally and strategically seek to defend itself, in the form of the provisional note.

The return to this literary form of expression also stands in a fundamental relationship to the archaic landscape, as suggested by a letter to Gretel Adorno of Spring 1932:

> These landscapes are surely the most inhospitable and most untouched of any habitable landscapes I have ever seen. It is difficult to convey a clear idea of them. If I finally succeed in doing so, I will not keep it a secret from you. For the time being [*Vorläufig*], I have not made very many notes with that in mind. On the other hand, I surprised myself by again employing the descriptive form of *One-Way Street* to deal with a number of subjects that are related to the most important topics of that book.[17]

Benjamin's provisional (*vorläufig*) observation raises the question: was Benjamin getting ahead of himself in the letter to Gretel Adorno in the spring of 1932? Yes and no. Just as he may not have foreseen at this point how so much of his writing in the coming years would make use of the miniature prose form, the sparse landscape that he mentions only sparsely might play a greater role in the formation of the weighty writing taking shape at the time of the letter than he admits. That he reports having noted "little" (*wenig*) specifically in regard to the Ibizan landscape makes "Downhill" all the more precious. Yet "little" in this context does not mean "unimportant" or "negligible." That Benjamin has taken few notes on the landscape does not mean that he has taken little note of it.[18]

Here it bears mentioning that the prose form that Benjamin takes up for these provisional notes is, in fact, first taken up on another rugged and sparsely populated island in the Mediterranean, Capri (and the adjacent coastal city of Naples), in a piece called "Naples" that was published in 1925. As biographers Howard Eiland and Michael Jennings note, his long stay in Capri "left indelible traces in Benjamin; his attempts to work through his experiences there in literary form remained a preoccupation to the end of his life."[19] One of the more salient impressions left by the island was a geologic one, and the geo-social observations that pervade "Naples" will remerge on Ibiza and "Downhill." Together with his coauthor and dedicatee of *One-Way Street*, Asja Lacis, in "Naples" Benjamin links up reflections on the geologic porosity of the region with sociological reflections on the region's cultures of provisionality: "As porous as this stone is the architecture. Building and action interpenetrate in the courtyards, arcades, and stairways. In everything, they preserve the scope to become a theater of new, unforeseen constellations. The stamp of the definitive is avoided. No situation appears intended forever, no figure asserts it 'thus and not otherwise'" (SW, 1:416; GS, 1.4:310). This geosocial "porosity," they continue, "results not only from the indolence of the southern artisan, but also, above all, from the passion for improvisation [*Improvisieren*], which demands that space and opportunity be preserved at any price" (SW, 1:416; GS, 1.4:310;). Martin Mittelmeier has argued that the porous, volcanic landscape of Naples facilitated the development of Adorno's thought.[20] A similar relation might be observed in Benjamin's encounter with the provisional geologies of this unsettled region and the development of the improvisational *Denkbild* form, which, as Gerhard Richter observes, "should be understood not simply as a prefabricated concept and fixed genre . . . but as the formal site for singular and unpredictable—but not arbitrary or facile—acts of conceptual creation."[21] This is also to note that the unsettled ground of the Ibizan thought images of 1932 is itself marked by a porosity in which one glimpses the encounter with porous structures in Naples, which itself is interpenetrated by Benjamin's childhood fascination with Berlin's subterranean spaces, a fascination that he will return to while on Ibiza in what will become his final thought image collection, *Berlin Childhood Around 1900*.

Nothing is immediately shocking on Ibiza; one cannot plunge into the crowd as though into a reservoir of electrical energy, because there is neither electricity nor crowds. But the island is shocking in an acoustic sense: the rocky ground is morbidly resonant. Its reverberations mark Benjamin's time. He writes of hearing his footsteps resonate during his hikes in the wilderness. Some say that the island is of volcanic origin and that the ground is composed in places of lava and tuff, but for Benjamin the local belief that there are everywhere graves underfoot is more compelling. "Small Notes About the Island— Here and there it sounds hollow when one takes a step. Could be that these are hollow places in the lava (if the island is in fact volcanic), but it is also claimed that these are graves" (GS, 4.1:454).

The piece of writing most sensitive to these reverberations is without question "Downhill." The closing scene in this miniature prose piece, a longer sentence describing the simultaneous ascent and descent of a mountain, contains a number of echoes of other's footsteps, ranging from Petrarch's summiting of Mont Ventoux in 1336, to the Easter Walk in Goethe's *Faust*, to Rousseau's solitary walks in *Les Rêveries du promeneur solitaire*. But in the search for literary echoes it should be recalled its earliest draft form "Downhill" bore the title "A Note on Proust" (*Proustnotiz*), and this note recalls yet another one, the "Notice of Death" (*Todesnachricht*) from *Berlin Childhood*, the collection of writings which he had also begun on Ibiza in 1932 under the name *A Berlin Chronicle*. This note too starts with someone startled by a morbid reverberation before closing with the revelation (and obscuration) of the death of a relative.

> The deja vu effect has often been described. But I wonder whether the term is actually well chosen, and whether the appropriate metaphor to the process would not be far better taken from the realm of acoustics. One ought to speak of events that reach us like an echo awakened by a call, a sound that seems to have been heard somewhere in the darkness of a past life. Accordingly, if we are not mistaken, the shock with which moments enter consciousness as if already lived, usually strikes us in the form of a sound. It is a word, a tapping, or a rustling that is endowed with the power to transport us into the cool tomb of long ago [*die kühle Gruft des einst*], from the vault of which the present seems to return only as an echo" (SW, 2.2:635; GS, 4.1:251–252).

Elsewhere in Benjamin's corpus, and even obliquely in this passage in the description of the past moment "stepping" into consciousness, a footstep is endowed with this power to awaken the past to a different trajectory and thereby give access to a future other than any one presently envisioned or otherwise perceived. Benjamin's walks in Ibiza are shocking in so far as they issue from an unarticulated and disarticulated ground: the cool crypt of long ago (*die Gruft des einst*) contains the echoes of the stones (*Stein*: an anagram of *einst*) that preserve the acoustic traces of the past in a kind of rock record, and accordingly they furnish the medium in which shock takes shape; the steps are shocking in so far as they recall those of his literary *precursors* (*Vorläufer*) and their mountain ascents; yet the precarious footsteps on the island and in particular in "Downhill" are perhaps the most shocking in so far as they are the *Vorläufer*, in the sense of *forerunners*, of those that carried him, in 1940, over the Pyrenees and into Spain in the unsuccessful attempt to escape from Europe that ended a day later with his suicide.

Downhill: Becoming-Mineral

"Downhill," and the ground on which it unfolds, acts not as a site of resistance to the impending disintegration that he notes in the letter to Sholem; it is the very scene of the disintegration of language, earth, and body. As a virtual anagram of *Grab* ("grave"), the title "Bergab" is one of these words whose echoes call us into the sepulcher of the past, or alternatively of the future: "the crypt of long ago" (*die Gruft des einst*) can also refer to the crypt of "some day" in the distant future, a crypt in whose anachronics and anagrammatics the present resounds only as an echo. The tremor that moves through Proust's body as he bends over in "Downhill" also moves through Benjamin's body as he walks across the island; and it also traverses the text of "Downhill." Shaken up and spaced out over the course of its articulation, *B-e-r-g-a-b*, like the hollow ground on which Benjamin treads, opens out into the haunting memory of a past or present burial (*begraben* means "to bury"; the past participle is identical) that itself cannot be interred, but returns cryptonymically an involuntarily, as parapraxis and paragram, and above all percussively, in the shattering of a word's integrity, the shuddering of a body as though undead, called forth from the Earth.

"Downhill"

> The word *"Erschütterung"* has been used ad nauseam. So perhaps something may be said in its defense. It will not distance itself from the sensual for an instant and will above all else cling to one fact: that shock leads to collapse. Do those who assure us after every theater premiere or every new book that they were "shattered" really wish to tell us that something inside them has collapsed? Ach, the phrase that stood firm beforehand will also stand firm afterward. How could they allow themselves the pause which is the precondition of collapse? No one has every felt this more clearly than Marcel Proust did when he learned of the death of his grandmother—an event which he found shattering but unreal, until the evening he burst into tears while taking off his shoes. Why? Because he bent down. In this way, the body is what rouses a profound pain; and it can serve no less to arouse profound thought. Both require solitude. Anyone who has climbed a mountain on his own and arrived at the top exhausted, and then turns to walk down again with steps that shatter his entire body—for such a person, time hangs loose, the partition walls inside him collapse, and he toddles through the rubble of the moment as if in a dream. Sometimes he tries to stop, but cannot. Who knows whether it is his thoughts that shatter him, or the roughness of the way? His body has become a kaleidoscope that at each step screens changing figures of truth. (SW, 2.2:592–593, translation modified)

"Bergab"

Das Wort Erschütterung hat man bis zum Überdruss vernommen. Da darf wohl etwas zu seiner Ehre gesagt werden. Es wird sich keinen Augenblick vom Sinnlichen entfernen und sich vor allem an das Eine halten: dass Erschütterung zum Einsturz führt. Wollen die, die uns

> bei jeder Premiere oder jeder Neuerscheinung ihrer Erschütterung versichern, nun sagen, etwas in ihnen sei eingestürzt? Ach, die Phrase, die vorher feststand, steht auch nachher fest. Wie könnten sie sich auch die Pause gönnen, auf die allein der Einsturz folgen kann. Nie hat sie einer deutlicher gespürt als Marcel Proust beim Tode der Großmutter, der erschütternd, aber gar nicht wirklich schien, bis ihm am Abend, da er sich die Schuhe auszieht, Tränen kommen. Warum? Weil er sich bückte. So ist der Körper gerad dem tiefen Schmerz Erwecker und kann es dem tiefen Denken nicht minder werden. Beides braucht die Einsamkeit. Wer einmal einsam einen Berg erstieg, erschöpft da oben ankam, um sodann mit Schritten, welche seinen ganzen Körperbau erschüttern, sich bergab zu wenden, dem lockert sich die Zeit, die Scheidewände in seinem Innern stürzen ein und durch den Schotter der Augenblicke trollt er wie im Traum. Manchmal versucht er stehen zu bleiben und kann es nicht. Wer weiß, ob es Gedanken sind, was ihn erschüttert, oder der rauhe Weg? Sein Körper ist ein Kaleidoskop geworden, das ihm bei jedem Schritte wechselnde Figuren der Wahrheit vorführt. (GS, 4.1:409)

Faced with an apparently unreceptive recipient, the account of the shudder in "Downhill" disarmingly incorporates the lack of affect into its arsenal; it arms the nonreaction, the absence of shock, with a nuclear reactivity. The shudder initiates collapse not despite but *because* it holds fast to a gradient—the gradient signaled by "Downhill"—no matter how imperceptible the decline might be. The apparent pause before the collapse will not have been a pause but a setting into motion of collapse. In this break, to which one platitudinously "treats" oneself, and in which the inconsequentiality of being shaken would register itself, a gradual countermovement to self-satisfaction and self-assurance amasses itself. In the phrase "that stood firm beforehand will also stand firm afterward" (*die vorher feststand, steht auch nachher fest*) a stutter that will have been audible only after the collapse it initiates, the *Einsturz*, staggers the sense of the sentence: *staggers* in the sense of arranging the shudder so that its implementation, its *Wirklichkeit*, is spread out across a variety of registers; *staggers* in the sense of distributing the opening phoneme of *Sturz* across the sentence; and *staggers* in the chronological sense that the shudder does not coincide with the collapse. In the syncopated order of *Shattering (Pause) Collapse*, Benjamin articulates a form of staggering that is articulated in the word *stagger*, which, unable to coincide with itself, is semantically staggered across a faltering of speech, feet that move with devious steps, vessels like Benjamin's off-course ship that move unsteadily from deviation to deviation. The link to the nautical, which occurs under the sign of a seasickness, nausea, tenuously emerges in the play of translation (where *zum Überdruss* is rendered "ad nauseum") and further exposes the lability of the original German text. But rather than producing nonsense, it is this very faltering of voice, foot, vessel, and earth that leads to sense.

From the reflection on the "shattering" experience of every artwork, Benjamin turns to Proust collapsing in grief, a year after his grandmother's death. This belated notice of death in Proust's sudden collapse flips a conventional temporality on its head by disclosing the present as the reverberation of a traumatic past or future event that cannot be said to be past, since it was never wholly present, a past or future that extends into a present itself suddenly deprived of its own contemporaneity. Likewise, there is a disproportionately consequential force of this brief text, which disturbs established sequentialities, whether graphematic, psychosomatic, chronological, or syntactic. Time and body hang loose, syncopal. Already the passage in Proust to which Benjamin refers itself concerns a chronic anachronism, "which so often prevents the calendar of facts from corresponding to the calendar of feelings," and which registers itself in the image of Proust's grandmother, whose death shook him up but did not register with him until over a year after her burial, when, upon bending over to take off his shoes, tears come.[22] The energy unleashed by Proust's shuddering body triggers the entrance of another shuddering body:

> Anyone who has once climbed a mountain alone, and arrived at the top exhausted, only to then proceed downhill with steps that shatter [*erschüttern*] his entire body, for such a person time hangs loose [*ihm lockert sich die Zeit*], the partition walls inside him collapse, and he toddles [*trollt*] through the debris of moments [*Schotter der Augenblicke*] as if in a dream. Sometimes he tries to stop, but cannot. Who knows whether it is his thoughts that shatter [*erschüttert*] him, or the roughness of the path? His body has become a kaleidoscope that at each step presents him with changing figures of truth" (SW, 2.2:592–593, translation modified; GS, 4.1:409).

The scene of his ascent of the mountain immediately flipping into a descent—a turn away from the tendency of mountain ascent narratives to linger contemplatively on the summit—is an image of nondirectionality that is reproduced on the level of every individual moment. Benjamin's description of the island to Gretel Adorno as the most "untouched" (*die unberührtesten*) takes on a new sense in this light: it suggests that the course of this walk describes a curve for which there could be no tangent. Which is to say that every moment is pivotal. These steps, presented in a writing both remarkably dense and remarkably sparse at the same time, shatter time into so much debris, so many

Sekundenbruchteile, fragments and fractals that blow up in time, as seen in the shift in the tense of the sentence describing the mountaineer summiting from the past tense (*erstieg . . . ankam*) to a lyric present (*erschüttern . . . sich wenden . . . lockert sich . . . stürzen ein . . . trollt*). The punctuated and punctual "once" (*einmal*) that initiated the description of the ascent swells out into the vast sparseness of an open-ended "sometimes" (*manchmal*). Time, out of joint, hangs loose. Asymptotic, the walker approaches the end without ever arriving.

Out of the shake-up that Benjamin initiates, nothing, not even the figure of the shake-up, emerges unshaken. The energy unleashed by the shudder goes so far as to dismember even the integrity of the shudder as a discrete term: the figure that would stand at the outset for an authentic aesthetic response, the possibility of being "moved," is itself subjected to its own shaking action in the course of the text, eventually giving way to the word *Schotter*, an aggregate designation for materials including "crumbled and crushed stone, isolated blocks found on flat fields, foundlings, related to *schutt* [debris, scree], *schütten* [to pour, to dump]."[23] Such an "irregular way"—whether traversing a talus slope, scree field, or asyndetic sentence—informs an erratic mountaineer's disjunctive temporal perception. From a scattering of the Earth to a shattering of perception by the uncoordinated, unsteady movement of a tottering pedestrian, Benjamin elaborates the geologization of experience in which "deep" thought emerges out of an embodied engagement with the deep time of a heavily weathered and stressed landscape.

A peripatetic body, whose plantigrade movement is resolved into discontinuous and unsteady steps, no longer aggregates the disparate debris of each moment into a landscape, no longer resolves the phrases into a syndetic whole. This unavailability of a conventional gait returns in the solitary mountaineer of "Downhill" who "toddles as though in a dream [*er trollt wie im Traum*]" recalls Benjamin's difficulty walking on the snow-covered sidewalks of Moscow: "*Ich muss mich trollen* [I have to walk like a toddler]" (GS, 4.1:318). Far from regressing into the tottering state of a toddler, though, the progressivity of "Downhill" consists in differentiating and disseminating the variety of "goings" within the conceptual and rhythmic monotony of an undifferentiated and undisturbed going. No form of the verb "to walk" (*gehen*) will occur once in this, the only existing literary record of Benjamin's numerous solitary walks in Ibiza. In the ongoing writing of his childhood—and it would be interesting to re-read *Berlin Childhood*, for example, with its crooked streets and subterranean trolls in terms of a kinesthetic imagination that does not yet avail itself of the monotonous form of walking described by *gehen*—the word "to go" falls out of circulation in Benjamin's lexicon. In order to access a spectrum of perception otherwise unavailable to consciousness, he moves toward accessing a stage of awareness and forms of locomotion that will never be anything more than *vorläufig*, provisional, which is to say: erratic, archipelagic. The *terrain erratique* of "Downhill," as mediated though similarly erratic fragments of language, shatters both landscape and body into a kaleidoscopic scree field traversed by a walking, talking crystal—a mineralized self not unlike the liquidated one that is the subject of Adorno's aesthetic experience. When not outright asyntactic, the prose of this sentence becomes asyndetic, marked by the omission of coordinating conjunctions, a syntax whose figurative recapitulation in "steps that shatter his entire build" presents a physiological stylistics that approximates that of an unlikely mountaineer but ultimately moves beyond a somaticization of language and earth.

Even though Benjamin begins "*Downhill*" by indexing shock to a human body and aesthetic experience, in the course of this brief text the horizon of shock opens out from the constricted one of an organism into the nonanthropic one of a prismatic body-kaleidoscope without a human face and yet many-facetted. This eseptate body approaches, with its self-shuddering footsteps, the uprising that closed *One-Way Street* in "To the Planetarium": "During the nightly annihilations of the last war the human frame was convulsed [*erschütterte den Gliederbau der Menschheit*] by a feeling akin to the bliss of the epileptic. And the revolts that followed it were mankind's first bid to bring the new body under control" (SW, 1:487; GS, 4.1:148).

But the tremors that ran through this collective body here lead to the collapse of an anthropotechnics, out of the rubble of which a solitary apparatus toddles downhill, having not yet reassembled itself according to any established, normalized ordering of motoric functions. The insistence on the essential solitude of this endeavor in "Downhill" would unduly isolate the shuddering mountaineer's body (*Körperbau*) from that revolting political body (*Gliederbau*) of "To the Planetarium," if Benjamin were not elsewhere insistent that even the most idiosyncratic gestures performed in solitude bear the mark of a historical conditioning. Not contained in the script of any social choreography, resistant to rehabilitation by any representational regime, and emancipated from civic strictures of traffic, this body cannot stand for anything, not even itself. Perhaps this is its politics.[24] *Sometimes he tries to stand and cannot.* Something has been set in motion in these lines, whose outcome cannot be determined.

Toward a Critical Climate Change

"Downhill" unfolds along the bending bodies and sloping mountains that approximate the ancient sense of "climate," derived from κλίνειν (*klinein*, "to lean," "to incline") and referring to the theory that the Earth sloped away from the equator, creating distinct meteorological zones. If "Downhill" is construed as a translation of "climate" in this antiquated sense, it bears observing that the trajectory of "Downhill" is far from uniform and that any model of climate extrapolated from it would be anything but stable. This bears mentioning given the notable passages in Benjamin's other texts from the thirties suggesting that he, as Tobias Menely argues, "nostalgically holds onto the promise of climatic continuity as a condition of historical self-awareness."[25] Drawing on the war-torn "landscape where nothing remained unchanged but the clouds" in "The Storyteller" and the "breath of the air that pervaded earlier days" that "caresses us as well" in "On the Concept of History," Menely finds that, in treating the air as an "archive of the other times that remain," Benjamin relies on "the periodicity of a stable climate" as a backdrop against which historical change can then be charted.[26] Although Benjamin's writings from this period certainly index social and political turmoil, there is little indication that they register the atmospheric turmoil that will come to be known as anthropogenic climate change. Starting from "Downhill," however, a different conclusion could be reached.

In the first place, it is precisely as an "archive of the other times that remain" that for Benjamin the atmosphere would be open, rather than resistant, to the thought of climate change. Doesn't the methane gas that pervaded pre-Holocene days and that now caresses us as well—having been captured in the Arctic in permafrost and the ice crystals that are now melting at an unprecedented rate—exemplify and affirm Benjamin's question from the second thesis on the philosophy of history that for Menely is taken to be evidence of a thinking *resistant* to climate change ("Doesn't the breath of the air that pervaded earlier days caress us as well?")? If so, for Benjamin the air would not be, strictly speaking, a figure of continuity as much as one of latent potency and as such *open* to the possibility of climate change. As theorists of climate change attempt to figure, in Menely's words, "this elusive atmosphere that bears the weight of history," we would be wise to not disregard Benjamin's figurations of air and atmospheres.[27] Benjamin's historical materialism will have critical purchase on the Anthropocene, understood as a moment of destratification, as I will elaborate in the epilogue. Indeed, Michael Levine regards Benjamin's above question and the entire second thesis as *"releasing something into the air"* rather than *"just letting something linger in the air"* in the mode of a stable archive.[28] But for all of the latent potencies of air in "On the Concept of History," Menely's observation holds that in Benjamin's writing air offers little indication of climactic irregularity. Not the atmosphere but the lithosphere, particularly where it appears as "the unarticulated earth" and "the unregulated earth," is the site where the past accretes irregularly and where a stable periodicity and periodization is shaken up. Unarticulated, irregular, rough, hollowed-out, and subject to shattering, the Earth is a medium of history for the historical materialist Benjamin. Since the air offers such scanty evidence of his knowledge (or unthought knowledge) of how historical change will have been registered in an atmospheric index, this chapter has implicitly argued that for Benjamin the Earth, not despite but because of its elusiveness and porosity, bears, in not bearing, the weight of history.

Epilogue: Dilapidated

On the wall there stood in chalk:
They want war.
The one who wrote it
Has already fallen.

—BERTOLT BRECHT

Perhaps one of the more formidable artistic responses to the Earth-magnitude events collected under the epithet of the Anthropocene takes the form of an epigrammatic call: *Help Us*. Installed in 2008 on a rooftop adjoining Pittsburgh's Carnegie Museum of Natural History and adapted from the heavily mediated spray-painted pleas on the roofs of New Orleans' flooded Ninth Ward in the wake of Hurricane Katrina, Mark Bradford's lapidary artwork consists of thousands of small white stones arranged into this 21 × 82 foot message that might be directed, one imagines, to Air Force One, a remote sensing satellite, or an extra-terrestrial intelligence, given that it was commissioned for a *Life on Mars* exhibition.[1] Just as the interpellant of this call extends beyond politicians and coast guard aircrews, the remediation and installation of this piece on the roof of a museum of natural history also massively scales up the appellant, from the black residents of the Lower Ninth Ward to the entire species, if not to the entire animal, vegetable, and mineral kingdoms. In its repositioned context, it might not be too immoderate to read *Help Us* as a call from the rocks themselves, on behalf of all things living and nonliving.[2] A similar amplification takes place in the call itself: the enlarged size of this text, its relocation hundreds of miles inland, many degrees of latitude north, and hundreds of feet above sea level transforms its context from hurricane flooding into a climate change–based sea-level rise more severe than most current projections—from what Tom Cohen and Mike Hill have called the "preemptive first strike in an undeclared U.S. civil war," namely Katrina, to "the generalized state of war" that Bruno Latour identifies as the Anthropocene, though the latter should also have critical purchase on the racial violence associated with the former.[3] That (at the time of writing) the museum's current special exhibition is "We are Nature: Living in the Anthropocene," reinforces a reading of *Help Us* as an epigraph for an Anthropocene in and as a time of war.

Although the remediations and transformations involved in Bradford's *Help Us* are notable, perhaps equally remarkable is its retrieval of the elementary situation of an ancient epigram form in which, to cite Geoffrey Hartman in another context, "the inscription calls to the passer by in the voice of the genius loci" and thereby "allows the landscape to speak directly, without the intervention of allegorical devices."[4] Hartman's definition of the ancient votive epigram, in a seminal essay on its reception by romantic nature poetry, could be said to pertain to the inscription mode of Bradford's *Help Us*, not to mention the works of other prominent contemporary artists whose work takes the form of outdoor inscriptions.[5] Here too is another transformation, that of the epigram's scope: for Bradford, who remediates this message on a natural history museum and with stone ("implying a problem at once archaic and enduring into the future," as art historian Katy Siegel observes), the role of the itinerant passerby in this tradition is not played by a bipedal wanderer as much as it is by an itinerant satellite or an alien geologist (fig. 5). (I happened upon it in 2012 while searching Google Maps for directions to the museum.) And although *Help Us* places its reader in question, something common to the epigraph, the epigram, and the epitaph at different points in their histories, it does so at a scale more common to stratigraphy. Thus, the transformations it enacts could be said to encompass both content and form: in the process of amplifying an individual's spray-painted plea into a natural-historical epigraph, it reimagines the epigraph (and arguably also the epigram) as a planetary form. What is more, the amplification of scale involved in this remediation—with respect to the size, its temporality, inferred authors and inferred audiences of *Help Us*—mirrors the expansion of "inscription" in contemporary geological usage to refer to human activity writ large in the lithosphere. In light of the questions of legibility raised by *Help Us*, the question of literature posed at the outset of this study—how the geochronological designation of the Anthropocene could furnish a frame in which literature and the literary would still be pertinent—could be productively reframed as a question of *reading* that would cut across both epigraphic and stratigraphic forms.

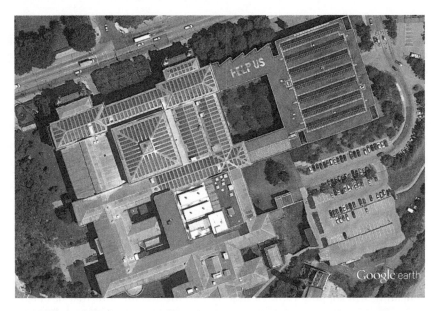

Figure 5. Screenshot of the Carnegie Museums of Pittsburgh (Google Earth Satellite View).

This mutual inclination of literature and history toward what could be termed "the lapidary" lies outside the focus of a study that until now has consisted exclusively in an inclination toward the erratic. Attending to this other inclination represents an occasion to address the apparent oversight of a significant scene of the mineral imaginary. Jeffrey Cohen's elaboration of two primary "propensities" of stone is useful in this regard: Cohen speaks of a "lapidary" propensity in which "stone captures narrative, stilling it within an enduring archive" and a "seismic" one in which "stone nonetheless remains active, on the move, and the stories held through its companionship likewise become unsettled."[6] This distinction is very useful even when, in light of the previous and current readings, it requires some revision. Although the seismic propensity encompasses an erratic one (as a mode of "stone on the move"), we have seen how the encounter with erratics has had a way of crystallizing narratives—albeit those of displacement, abandonment, and adventure—in an archival mode more commonly associated with the lapidary: the captivating metals of *Rune Mountain* and the trope of the cold heart that it conveys; the stationary granite blocks whose errant trajectories inform Goethe's *Wilhelm Meister's Journeyman Years*; the characteristic stillness of Stifter's *Many-Colored Stones*. The lapidary propensity is also evident in the epigram and similar literary traditions in the inscriptive mode, but in the Anthropocene the lapidary too has become a site of vulnerability and duress more commonly associated with seismic events, as the mediated images of *Help Us* and their sculptural remediation illustrates.

In modernity any clear distinction between the lapidary and the seismic becomes muddled, and this movement in turn recalls an earlier moment of unsettling that geological discoveries provoked for the literary arts. As Charles Lyell wrote in *Principles of Geology*, "But it is time that the geologist should in some degree overcome those first and natural impressions which induced the poets of old to select the rock as the emblem of firmness, the sea as the image of inconstancy."[7]

It might be time today to overwrite other impressions: those impressions that induced poets to associate the lapidary with endurance, and more fundamentally those impressions that induced poets and geologists alike to associate epigraphic and stratigraphic forms with lasting impressions. A poet writing in the lapidary style and his incisive commentator, Bertolt Brecht and Walter Benjamin, will support a line of inquiry into the seismic shift that has been underway since German romanticism and *Rune Mountain*. Although Brecht's turn toward the epigram form can be read in the context of a contemporary geological and literary inclination toward the lapidary, what is more pertinent is how his unsettled inscriptions question the perceived affordances of the lapidary (durability, archivability), whose behavior is increasingly erratic. By way of, or rather in lieu of, an epilogue to this study of the erratic, I offer an evanescent epilith, that is to say, some reflections on the shifting relations between writing and stone as mediated by a common dilapidation.

The Anthropocene as Epigram

One of the ironies of the present is that publications on the Anthropocene by the International Commission of Stratigraphy's Anthropocene Working Group conceptualize the Earth—in terms of the litho-, chemo-, biostratigraphic "signal," "marker," or "signature" inscribed in the geologic "record"—in terms more commonly associated with literary theory.[8] As Tobias Menely and Jesse Oak Taylor observe, although the working group makes two distinct claims about the Anthropocene—a geophysical claim about an altered Earth system and a stratigraphic claim about a unique interval in sediment deposits and ice cores—formal recognition of the Anthropocene in the geological community derives from a stratigraphic prerogative premised on a "semiotic claim about the clarity of a 'signature' recorded in a lithostratigraphic archive."[9] Literary theorists have embraced both claims, in particular the stratigraphers' expansion of signal, signature, and inscription to encompass apparently nonsignifying traces. For Tobias Boes the collapse of the distinction between human history and natural history in the Anthropocene casts the planet as a multidimensional *codex*—recalling the Medieval "book of nature" topos as well as the nineteenth-century trope of the "rock record"—of mundane rather than divine messages: "We have inscribed ourselves in its pages through our CO_2 emissions, our nuclear tests, our strip mines, and our general waste of the Earth's resources."[10] Momentarily setting aside the very fraught question of this elite "we" with whom Boes identifies, and what such a rhetorical gesture facilitates and occludes, for the moment I am interested in how the description of atmospheric and lithospheric transformation as a form of "textual expression" retrieves and reworks the metaphor of the book of nature; moreover, as Boes points out, such emissions and impressions would fulfill the "expressive" and "poetic" dimensions of language as defined by Roman Jakobson.[11]

Boes (following studies by Wai Chee Dimock, Jesse Oak Taylor, and Adam Trexler) turns toward the novel as the form that promises to salvage experience of the world in the Anthropocene; but the anthropogenic impressions left on the planet and in the Earth system, though epic in scale, often resist legibility and recognizable experience for future as well as present readers, and in this way they point to an aspect of the present that challenges narrativizing impulses traditionally associated with the novel. There are also empirical grounds for the unsuitability of the book, and consequently the novel, to account for a geologic legacy. "A *Lagerstätte* of books with preserved pages, somehow left flexible and uncarbonized by burial, is hard to envisage," observes Jan Zalasiewicz, chair of the Anthropocene Working Group.[12] Moreover, the idea of the book, "which always refers to a natural totality," as Derrida remarks in *On Grammatology*, also makes this metaphor less than apt today.[13] Following Derrida's dictum on the end of the book, Bronislaw Szerszynski reads the fate of the book together with the fate of the human, with the latter operating as a "destratifying force" and thereby as the book of nature's "condition of *im*possibility."[14] Where human activity manifests itself as igneous "extrusive and intrusive formations" that break the stratigraphic "logic of superposition," it represents a "falsification of the original text of the great stone book of nature" and could therefore be accommodated only by a much more volatile volume.[15] This conclusion is sound, even when the potential for the falsification, let alone the postulation of an original book of nature that would not already be a palimpsest, is as untenable for modern geology as it is for a grammatology. The shortcomings of the book and above all the novel, with respect to the literary imaginary of the Anthropocene, come from elsewhere. Given that the form of the novel, in Rüdiger Campe's observation, is theorized from the beginning "as a form of life" that unites old European poetics' and modern aesthetics' concerns with "lively rendering," "vivification," and the "art of life," the ability of this form to elaborate the stratigraphic imaginary of the Anthropocene can be put into question where it is furnished, as is often the case (see *The Earth After Us*), by an inhuman gaze from the distant future that regards the Earth as a graveyard containing relics of long-dead organisms.[16]

If it seems contradictory, at first glance, that the present study has included longer considerations of two novels, Goethe's *Journeyman Years* and Stifter's *Indian Summer* that strive to make sense of the world through the lively rendering of protagonists' life stories of biological and intellectual development, it is worth recalling that in both those cases aberrant stones perturb those biographical narratives, and the ability of narrative to render a well-plotted world, to the breaking point. An extreme case, if somewhat speculative, is offered by the fate of Goethe's *Cosmic Novel*, whose abandonment shortly after conception is curiously coterminous with his discovery of the granite foundlings that complicated the novel's putative reliance on the unshakeable granite bedrock of "On Granite" for its foundation. Although they achieve a higher level of completion, the novels *Journeyman Years* and *Indian Summer* are nevertheless notable for their periodic abandonment of narrative in favor of aphorism, prolix descriptions, and interpolated lyric. These postnarrative stagings of an "unplottable world," to use Samuel Frederick's fitting phrase for "a world that will no longer fit into the coherent categories we have inherited to make sense of it," are informed by their author's experiences of living on an unplottable planet.[17] Already for Goethe and Stifter it is increasingly uncertain, in the wake of the geological discoveries mediated and at times premediated in these books, that the ideal rendering of these worlds is necessarily a novelistic one.

Although the unreliable earth with which with the narratives in the present study contend is not yet that of the Anthropocene, they nonetheless are already beset by formal challenges greatly exacerbated by the Anthropocene imaginary. Indeed, rather than the novel, codex, or epic, some of the more prominent elaborations of the Anthropocene present it in terms of anthropogenic contributions to the planet's stratigraphy that suggest smaller forms such as the epigram and epitaph. As Jeremy Davies argues in *The Birth of the Anthropocene*, the thought experiment responsible for the stratigraphic Anthropocene "tries to understand the present by imagining its geologic traces," and such a corresponding commemorative faculty might be approximated more closely by the funereal epigram's memorializing impulses than the novel and its vivifying impulses. Moreover, Davies gestures toward the "concision and memorability" associated with this form in the epitaphic "Obituary for the Holocene" that closes *The Birth of the Anthropocene*.[18] (Samuel Frederick's definition of the postnarrative text, based on his reading of *Indian Summer*, as "a narrative seeking to escape from temporality into a utopian space *after the end of* any story—*after the end of narrative time*" seems remarkably pertinent for these retrospective accounts of the Anthropocene.[19]) Likewise, in *The Earth After Us*, Zalasiewicz casts the enduring trace of the human in what could nominally be described as an epigrammatic mode. As Zalasiewicz observes, "The strata of the Earth are graveyards, the burial places of relics from which long-dead organisms can be recreated. Visiting aliens will certainly be aware . . . that things of this sort are possible."[20] These more-than-human inscriptions and these extraterrestrial stratigraphers, albeit greatly amplifying those challenges to the epigrammatic tradition emerging already in the eighteenth century (illegibility and uncertain audience), are nevertheless contiguous with Geoffrey Hartman's identification of the epigram's function as "a call from a monument in the landscape or from the landscape itself."[21] Similarly, for Zalasiewicz's alien stratigraphers, whose task is "to find the message left by the human race" in sediment and rocks and "then to decipher it,"[22] the strata of the Earth are forecast to possess legibility that recalls John MacKay's *Inscription and Modernity* and its accommodating definition of epigram as both a "literary norm" and "the practice out of which that norm emerged: specifically, sepulchral and other commemorative kinds of place-marking, such as burial markers along the road, votive sites, memorial carvings, and the like."[23] It is this two-fold definition that facilitates and underwrites the construal of the Anthropocene as epigram.

Inscription and Geomodernity

An assessment of the potential legibility of these stratigraphic marks is afforded by MacKay's definition of epigram in *Inscription and Modernity*, given the significant if inadvertent commemorative function that stratigraphy is imagined as playing today. Of the four features constitutive of the inscriptive mode for MacKay—appellation, projection, corporeality, and legitimation—at least three could be said to persist in some way or another in the stratigraphic inscriptions associated with the Anthropocene. To what extent these actively summon an audience is questionable, although Zalasiewicz is not alone in imputing a wished-for reader to your and my lithic legacy when he writes that the anthropogenic rocks and strata would serve as the conveyance of "a final message, that of your own brief existence, into the next geological era."[24]

Notably, the uncertain audience and the radical illegibility of *these* final messages partake in the tendency, already present in nineteenth-century inscriptive poetry, to convey a "sense for a life (in nature) so hidden, retired, or anonymous that it is perceived only with difficulty."[25] Geoffrey Hartman (cited by MacKay) here refers to the "artless tale" of Thomas Gray's *Elegy Written in a Country Churchyard*, but it requires little imagination to extend this hidden and anonymous sense of life to the artless geochemical anomalies now present in sediment and ice cores. When traceable to sites of origination—such as the "widespread anthropogenic signal" produced by colonial metallurgy in Peru and Bolivia in the sixteenth century[26]—such signals intimate a rudimentary "articulation of space" with which MacKay primarily defines "projection," a key feature of the epigrammatic mode.[27] These durable signals in the cryosphere indicate the vast dimensions of the systems responsible for their deposition, and, in indicating these systems, they—whether as information-bearing "signal" or as "signature" indicating an absent hand, appendage or terraforming assemblage—achieve a measure of corporeality that extends beyond their obvious physicality. Just as the inscriptive mode common to epigrammatic and locodescriptive poetry acknowledges "some (immensely mediated) material substrate on which it depends," as MacKay's definition of corporeality runs, so too does the stratigraphic Anthropocene regard the planet, in the words of Tobias Boes, as "a medium for the storage and recursive transmission" of such "human-generated messages."[28] One could contend that the distinction between the mark and its material substrate no longer obtains in the case of such inscriptions (after all, the ice core *is* the Earth), but again, as MacKay points out, the unreliability of this division is already a feature of the postmodern aesthetic and "the erasure of any ontological difference between the representing 'medium' and the referent."[29]

The apprehension of stratigraphy as so many "messages" may give readers pause. It is one thing to argue that Mark Bradford's *Help Us* or even Carl Sagan's gold-anodized aluminum plaques, engraved with images of two humans and symbols pertaining to the Pioneer 10 spacecraft to which it was affixed and which in 1983 became the first manufactured object to depart the solar system, gather together "the writer, the wished-for reader, and the commemorated object" into the circuit that for MacKay defines the inscription's atmospheric function.[30] It would seem to be another thing entirely to argue that the novel stratal intervals of the Anthropocene could be read along these lines. Is there any plausible sense in which technofossils and the anthrostratum currently being deposited will actively summon an audience along the lines of the classical epitaph's imperative apostrophe (*siste, viator!*) or its reiteration in a well-known romantic epigram (*Halt, Traveler!*)? This seems highly improbable. Like MacKay observes with respect to his reflection on the persistence of a tradition of inscriptive poetry in commercial signage, the foregoing and following remarks on these stratigraphic inscriptions offer a kind of thought experiment rather than literary analysis. And yet, with respect to the question of an audience summoned by these inscriptions, the geologist Zalasiewicz is not alone in imagining his book as offering practical advice for optimizing "the message that we leave for far-future explorers to discover," nor is he alone in envisioning, in the opening vignette, an extraterrestrial geological community summoned by the Earth's stratigraphic anomalies.[31] Even if you affirm Jennifer Wenzel's riposte to Zalasiewicz's invitation in the preface of *The Earth After Us* to "read on" if you desire "to adorn some museum of the far future"[32] (Wenzel: "I do not read so that some other species might read about me in the deep future"[33]), and even when traces of these cataclysmic quasi-anthropogenic events in the rock register are inadvertent, and even when they have no single author, and even when any collective, let alone any individual, cannot alter the radioactive signatures already laid down, it seems prudent to acknowledge "that the Earth is a communicative object itself," as media theorist Jussi Parikka writes.[34] A future reader, no matter how far removed from any conventional image of a reader, cannot help but be anticipated, if not actively summoned, by anthropogenic contributions to the strata machine.

The question of posterity posed by these arguably inadvertent messages falls along a line of inquiry initiated by the poem taken by both MacKay and Hartman to be the exemplar of romantic inscription poetry. *Remember the old Man, and what he was*: in Wordsworth's line from "Michael: A Pastoral Poem," MacKay draws attention to the curious use of the relative pronoun *what* where we might expect *who*, which he takes to suggest "that the sturdiest basis for community and posterity emerging out of poetic inscription might rest on this radical appeal to a common materiality, what Paul Fry calls our 'mineral community.'"[35]

In a semiotic regime restricted to literature, such a community would be, as MacKay goes on to argue, strictly unrepresentable. But in the vastly expanded horizons both of semiotics and of posterity, a mineral community might become manifest in earthly inscriptions, inscriptions that nevertheless probably withhold any utopian hope that might have resided in such a common sense. As Wordsworth's well-chosen *what* and MacKay's careful commentary suggest, the basis of posterity is not a coming mineral community but rather a minerality that humans already share with many living and nonliving things, although it has only become more evident with the advent of the formations associated with the Anthropocene. Theorizing "our" lithic legacy would seem to stand to be informed by theories of poetic inscription—and, perhaps surprisingly, all the more so when that legacy and that theorizing can no longer be ascribed to a single species.

To add the stratigraphic records of the Anthropocene to the series of inscriptive markings considered in *Inscription and Modernity* (the epitaph, the modern inscription proper, and the commercial sign) would have to entail the registration of those "shifts in consciousness and sense of agency" that for MacKay correspond to the emergence of each of these modes of inscription.[36] This is indeed the case. Questions of agency have become enormously fraught in the Anthropocene, and this is not at all unrelated to the questions of reading and writing explored earlier. Even as early accounts of the Anthropocene relied on the human as the very "metaphysical centerpiece" that had been effectively displaced in postmodern thought, today most accounts concur with Jeremy Davies's recent observation: "Recognition of the Anthropocene does not . . . mean asserting that the whole world is now subordinated to human agency."[37] The strata currently being laid down, ripped out, or otherwise (re)configured cannot be solely traced back to the miners, frackers, drillers, manufacturers, consumers, investors, financial speculators, ideologues, and other drivers of fossil capitalism. To the contrary, as Davies explicitly argues, "Colorado beetles and chlorine atoms, like gray squirrels and kudzu, are among the powers of the Anthropocene."[38] Accordingly, the stratigraphic signatures in sediments and ice that will be putatively legible to alien geologists include the records left by marine transgression, mineral diversification, proliferation of invasive species as well as rapid species hybridizations. In Davies's estimation, such geologists of the future "could identify the present days as a significant turning point while learning very little about the part played in it by . . . *Homo Sapiens.*"[39] This new epoch, Davies argues, "emerges through a particular configuration of ecological agencies that has, among other things, temporarily multiplied the effectiveness of *some* human actors."[40] Beyond critical race theorists' important observation that the imputation of Anthropocene causes and effects to all of humanity (e.g., "our nuclear tests, our strip mines, and our general waste of the Earth's resources") effectively papers over the settler colonial histories driving many of these events, Davies argues that the geologic legacy of colonialism, imperialism, and capitalism is not even exclusive to the humans responsible for those histories but must also include all manner of more-than-human agents and processes.

If reading the Anthropocene as epigram (or even as epitaph) sounds alarmingly defeatist—the stratigraphic Anthropocene as planetary necropolitics, as premediation of environmental justice movements' capitulation—Davies allays some of these concerns when he suggests that the "alien perspective" of a retrospective stratigraphy is one "in which plastics, grasses, humans, plankton, and carbon monoxide molecules were all bundled together," and thus it is not biological extinction that is seen as being inevitable as much as the end of an anthropocentric gaze.[41] Ecocritical readings of epigraphic writing might take note from speculative stratigraphy's decentering of the human, in particular as it might help us, like Bradford's remediated *Help Us*, evoke a common vulnerability and accordingly imagine forms of survival that accommodate a more-than-human future.

Intemperate Climate, Intemperate Commentary

The question that animates *Inscription and Modernity* also animates this inquiry, namely: "How does all this help us to read poetry?" and for that matter prose.[42] If putatively anthropogenic inscriptions can be construed along the lines of a multispecies epigrammatic object, can epigrams be construed as more-than-human messages? Here we could do much worse than take up Eileen A. Joy's advocacy (on the most lapidary of social media platforms, Twitter) for a literary criticism that would "multiply and thicken a text's sentient, bottomless reality" and which "would be better described as a commentary that seeks to open and not close a text's possible 'signatures.'"[43] One of the many challenges for such a commentary is to attend to these multiple signatures—perhaps in the form of what Tobias Boes and Kate Marshall term *ecodiegesis*—rather than the traditionally ecomimetic nature poetry that required a poet's authoritative voice to render the nonhuman world audible and legible.[44] Perhaps no writer has advocated more vocally for commentary, nor reflected more acutely on its challenges in a time marked by catastrophe, than Walter Benjamin.[45] His reflections on the commentary form, particularly those made with respect to a series of epigrams written during the *Weltuntergang* that could arguably be extended to encompass the planetary catastrophes of the Anthropocene, will provide some provisional answers to the question of how all of this might help us to read literature, as well as its necessary correlate: *How does all this disallow us to read literature*? The extent to which this commentary now appears ill-suited to grasp the scale of that catastrophe is nevertheless an indication of its remarkable prescience. It is acutely aware, without ever saying so much, of how our inattentiveness to literary texts' possible signatures constitutes a great derangement, to take Amitav Ghosh's phrase, one that has become evident in the ongoing alteration of the Earth system and the exposure of all those scenes of reading and writing that were previously sheltered from the thought of their planetary contexts, even when those texts' occasional appeals to a mineral community uncannily anticipated the Anthropocenic inscriptions that we are confronted with today.

Benjamin's 1939 commentary on the poems of Bertolt Brecht poses the challenge of reading during a terminal phase of the world that turns out to have been the terminal phase of the Holocene. Following Jeremy Davies's appeal to understand the birth of the Anthropocene by charting the death of the Holocene, Benjamin's commentary can be read not only as an attempt to take the measure of a declining epoch but also an emerging one. His avowedly *intemperate* commentary exposes the becoming-intemperate of texts during the interval in which a geologic epoch characterized by relative climatic regularity and periodicity ceases to exist. In an era where the passing of time has become highly irregular, Benjamin's immoderate attempt to salvage the classical practice of commentary is conspicuously inadequate to grasp a poetry that is itself characterized by irregular rhythms and that, as we will see, takes as its subject the asymmetrical warfare waged by an unlawful combatant against a rogue state. His obstinacy is all the better to measure the erraticism of his time and its literary objects.

More recently, Timothy Clark has repeatedly called attention to a widespread blindness to questions of scale in dominant modes of reading.[46] The urgency in addressing this blindness, according to an article by Clark in 2012, derives from the degree to which climate change "disrupts the scale at which one must think, skews categories of internal and external, and resists inherited closed economies of accounting or explanation" including liberal, individualist, and anthropocentric traditions in political thought and literary criticism.[47] Even though his argument has been reiterated and reinforced on several occasions, including in Derek Woods's "Scale Critique for the Anthropocene" (2014), Clark's own *Ecocriticism on the Edge* (2016) and more recently in Benjamin Morgan's "Scale and Form" (2017), Clark's conclusion remains somewhat striking. Even though most thinking about literature and culture has taken place on a human or national scale, one regarded as unable to register or otherwise respond to environmental emergencies of the long-term and large-scale, Clark concludes that an adequate response of literary criticism to climate change will not entail the invention of a new method of reading, "for its most prominent effect is of a derangement of scales that is also an implosion of intellectual competences."[48] For Clark this implosion stems from what he in *Ecocriticism on the Edge* calls the "emergent unreadability" of literary texts in an era of climate change.[49] He illustrates this state of affairs in a pointed reading of Gary Snyder's "Mid-August at Sourdough Mountain Lookout," a poem about Snyder's stint as a fire-watcher in the Cascades that had been read by many as a "green" poem but which implicitly endorses and celebrates a practice of fire suppression that has since been found to be ecologically insensitive if not also destructive. Clark refers to this shifting assessment of the poem as its "emergent sense," in that it "exceeds that of the situation in which it occurred" and moreover where "that situation is being reconceptualized as a context that must now also include the present and an uncertain future."[50]

The implosion of critical competence entailed by an emergent sense happens to be the very difficulty addressed by Walter Benjamin's 1939 commentary. "The Form of the Commentary," the opening theoretical section, is animated by Benjamin's concern with the emerging unreadability of lyric: "How, in this day and age, can one read lyric poetry at all?"[51] This difficulty may be common to modernist literature, but it is also particular to a time and place marked by the threat of an immense scale of destruction. Benjamin attempts to overcome this difficulty through the elaboration of a contradiction that persists as a critical impasse today: a crisis of reading whose solution does not involve the development of a novel method but rather the explosion of the historical and generic conventions of traditional forms of reading. Benjamin's awkward but effective deflection of the difficulty of reading takes place in his application of commentary—"a form that is both archaic and authoritarian" and that "takes for granted the classical status of a work under discussion"—to a "body of poetry that not only has nothing archaic about it but defies what is recognized as authority today" (SW, 4:215; GS, 2.2:540). Regarding his somewhat deranged attempt to read contemporary lyric as if it were a classical text, Benjamin writes that his encouragement is summoned from the desperation that is the source of all audacity today:

> the knowledge that tomorrow could bring destruction on such a scale that yesterday's texts and creations might seem as distant from us as centuries-old artifacts. (SW, 4:215)

> dass nämlich schon der kommende Tag Vernichtungen von so riesigem Ausmaß bringen kann, daß wir von gestrigen Texten und Produktionen wie durch Jahrhunderte uns geschieden sehen. (GS, 2.2:540)

The derangement of temporality in a time of war—not the modernist chronophilia of Proust and Woolf but the chronophagy of an alternative, moribund modernity—offers a provisional legitimization of his undertaking to write this intemperate commentary. The image that he uses to illustrate the unsuitability of the commentary is striking:

> (The commentary which is such a tight fit today may hang like classical drapery by tomorrow. Wherever it might seem almost indecently precise now, it could be full of mystery tomorrow.) (SW, 4:215)

(Der Kommentar, der heute noch zu prall ansitzt, kann morgen schon klassische Falten werfen. Wo seine Präzision fast indezent wirken könnte, kann morgen das Geheimnis sich retabliert haben.) (GS, 2.2:540)

The tentative solution to the problem of reading in such times, as Benjamin writes and as Clark reiterates, is not to find an "apt" mode of reading but instead to affirm the virtual ineptitude of any reading. Although the comic effect of an increasingly baggy garment tempers the declensionist story of the end of the world, it remains the case that, like the fairy tale of the water nymph Melusine that he had long planned to address, Benjamin is charting a history of reduction, dissipation, and recession of literary objects—but also their increasingly erratic behavior.[52]

Benjamin is charting an "age of asymmetry" in which he himself is enmeshed. The phrase comes from Timothy Morton, who argues that the asymmetrical relationship between science and the objects of scientific knowledge is one of the defining characteristics of the present: "We know more than ever before what things are, how they work, how to manipulate them. Yet for this very reason, things become more, rather than less, strange. Increasing science is not increasing demystification."[53] Benjamin is charting a related asymmetry, albeit of literary objects. For Benjamin reading entails becoming attuned to the capacity of literary texts to dramatically deviate from the commentarial techniques to which readers have subjected them. Consequently, techniques of reading obtain a peculiar place in the diagnosis and prognosis of the coming destruction through a demonstration of their very diminishment and debilitation: if the volatility of the literary object indexes the magnitude of destruction, then the attendant debilitation of literary criticism, philology, and theory offers an additional measure of this destruction. Critical impotence reflects the increasing potency of writing. The significance of this position increases not despite but because of his commentary's inability to address the planetary crisis that was emerging alongside the political and historical crisis to which it most directly responds.

After the Lapidary

For Benjamin the shifting valence of the lapidary indexes planetary change. As he turns toward stone and the style of writing developed for stone, he finds only ephemera. In this way his remarks might be more aligned with a destratifying Anthropocene than they are of the Holocene's consolidated strata, with the former taken as a moment in which "what is in the ascendant is thus not the durable world of things made by *Homo faber*, but impermanence and change" consisting in "the fluxes and flows of substances such as CO_2, SO_2 and NO; the migrations of species; the transformation of ecological communities; accelerated erosion and denudation."[54] More recently Jennifer Wenzel has also elaborated a Benjaminian critique of stratigraphic logic based on an analogy between stratigraphy and "the kind of thinking Benjamin derided as historicism, where the past is safely past, neatly buried under the present in smooth and legible layers."[55] The untenability of the lapidary style that Benjamin documents might then correspond to the increasing untenability of the Holocene and moreover the untenability of a stratigraphic logic of superposition to account for historical change in a geological epoch characterized by destratification and resource extraction.[56] The writing that persists across time and space, Benjamin suggests, consists of erratic inscriptions without a reliable ground.

Although Benjamin's Brecht commentary ranges from the 1927 *Hauspostille* ("Household Messenger") to the 1939 *Svendborger Gedichte* ("Svendborg Poems"), it is in relation to the "Deutsche Kriegsfibel" or "German War Primer" poems of the latter that the lapidary comes into question. However, as Benjamin indicated already in the opening lines, the object of his commentary is less the veneration of antiquity and far more the charting of anachronism and dilapidation. The deranged temporality of his historical situation becomes evident, if not yet in the transformation of the lithosphere then in the transformations of the lapidary style. Benjamin's commentary begins, "The 'Kriegsfibel' is written in a 'lapidary' style. The word 'lapidary' comes from the Latin *lapis*, meaning 'stone,' and refers to the style which was developed for inscriptions [*Inschriften*]" (SW, 4:240; GS, 2.2:563). At this point Benjamin switches to the past tense: "The most important characteristic of this style was brevity [*die Kürze*]. This resulted, first, from the difficulty of inscribing words in stone, and, second, from an awareness that anyone addressing subsequent generations ought not to waste words" (SW, 4:240; GS, 2.2:563). The apparent obsolescence of the poem's style, already hinted at by the tense change, is then made explicit: "Since the natural, material conditions of the lapidary style do not apply [*wegfällt*] to these poems, one is entitled to ask what corresponding factors are present here. How can the inscription style of these poems be explained? [*Wie begründet sich der Inschriftenstil dieser Gedichte?*]" (SW, 4:240; GS, 2.2:563).

To be more precise, Benjamin is asking how the lapidary style of these poems can be *grounded*. The lapidary can no longer be grounded in stone. In adopting a style that has outlasted the material conditions for which it was developed, Brecht's lyric appears anachronistic and unfathomable. Its brevity is immoderate, its ground insufficient. A number of such contradictions will turn out to animate the poem that for Benjamin is exemplary of this style, the poem whose subject is mediated by stone dust, the poem which hints at the answer to the question of the ground and as such whose first line "could preface each of the poems in the 'Kriegsfibel'":

> On the wall there stood in chalk:
> They want war.
> The one who wrote it
> Has already fallen.
>
> (SW, 4:240)

> Auf der Mauer stand mit Kreide:
> Sie wollen den Krieg.
> Der es geschrieben hat
> Ist schon gefallen.
>
> (GS, 2.2:563)

This poem in particular and "The War Primer" poems in general, a collection which would become the basis of Brecht's better-known 1955 collection of "photo-epigrams" also called *The War Primer*, can be situated in close proximity to the epigram tradition. For one, August Oehler's German translation of the Greek Anthology of epigrams appeared in 1920 with the same publisher as Brecht's 1927 *Hauspostille*, the Berlin Propyläen-Verlag, and Brecht's copy contains annotations from the year of publication. As is evident from several well-documented journal entries, Brecht will return to this volume in earnest two years after the publication of *Svendborg Poems*, but as Christian Wagenknecht surmises, Brecht's conversations with Benjamin about Baudelaire in 1938 may have well involved Oehler's book.[57] Aside from Brecht's documented reception of ancient epigram traditions, his familiarity with the epigram form is made apparent when the previously cited poem is read alongside an account of the epigram provided by one of Germany's premier theorists of this form, Gotthold Ephraim Lessing: "The authentic inscription is not to be thought without that whereon it stands, or might stand. Both together make the whole from which arises the impression which, in an ordinary manner of speaking, we ascribe to the inscription alone. First, some object of sense which arouses our curiosity; and then the account of this same object, which satisfies that curiosity."[58]

From the opening line of Brecht's poem, which calls attention to the "object of sense" (*sinnlicher Gegenstand*) on which the message "stands" (*steht*) and from which it cannot be thought apart, to the cryptic "message" (*Aufschrift*) in the second line, and to which we solely ascribe the "impression"

(*Eindruck*) that the message produces, and finally to the explanatory "account" (*Nachricht*) of the object in the third and fourth lines, Brecht recovers to the letter this elemental epigrammatic situation of the inscription as elaborated by Lessing. And yet Brecht subverts this tradition in equal measure, as Benjamin carefully points out in his commentary. In turn, Benjamin extends what is intemperate in Brecht's appropriation of the epigrammatic mode: a poetic situation where the most ephemeral becomes the most durable, where brevity obtains duration, and where the lapidary style is thoroughly dilapidated:

> The first line of the poem could preface each of the poems in the "Kriegsfibel." Their inscriptions are made not for stone, like those of the Romans, but, like those of outlawed fighters, for palisades.
> Accordingly, the character of the "Kriegsfibel" can be seen as arising from a unique contradiction: words which through their poetic form will conceivably survive the coming apocalypse preserve the gesture of a message hastily scrawled on a fence by someone fleeing his enemies. The extraordinary artistic achievement of these sentences composed of primitive words resides in this contradiction. A proletarian at the mercy of the rain and Gestapo agents scribbles some words on a wall with chalk, and the poet invests them with Horace's *aere perennius*. (SW, 4:240)

> Die Anfangszeile dieses Gedichts könnte jedem der "Kriegsfibel" beigegeben werden. Ihre Inschriften sind nicht wie die der Römer für Stein gemacht sondern wie die der illegalen Kämpfer für Palisaden.
> Hiernach darf der Charakter der *Kriegsfibel* in einem einzigen Widerspruch erblickt werden: in Worten, denen ihrer poetischen Form nach zugemutet sind, den kommenden Weltuntergang zu überdauern, ist die Gebärde der Aufschrift auf einem Bretterzaun festgehalten, die der Verfolgte mit fliegender Hast hinwirft. In diesem Widerspruch stellt sich die außerordentlich artistische Leistung dieser aus primitiven Worten gebauten Sätze dar. Der Dichter belehnt mit dem Horazischen aere perennius das, was, dem Regen und den Agenten der Gestapo preisgegeben, ein Proletarier mit Kreide an eine Mauer warf. (GS, 2.2:564)

Although only amounting to seven sentences, Benjamin's commentary on this poem is inordinately sensitive to these lines and the fraught situation they describe. At the same time his remarks and the scale they adopt no longer fit the poem that can be read today as an allegory of the Anthropocene: *a poem about reading the traces that outlast the one(s) that left them.* To do so would be to commit an immoderate anachronism—not the least of which, it entails expanding the coordinates of the "lapidary style" to encompass multispecies lithostratigraphic inscriptions—but Timothy Clark's argument that texts are legible "in times and ways that are not those of their own day" is also the wager that Benjamin's stakes in his initial remarks on the commentary form.[59] Benjamin's wager on his commentary's unsuitability is won, posthumously, by missing the mark. And if the poem is read as an allegory of writing in the Anthropocene, Benjamin's commentary might anachronistically furnish elements of a theory for the same.

The Afterlife of the Lapidary

Although Benjamin's commentary on this poem begins by situating Brecht's lyric in a postlapidary period, it will conclude by retrieving the lapidary as an art of survival. Literature would then be after the lapidary in two contradictory senses of "after": temporal and mimetic. Although stone is increasingly unavailable for this literature—because the style of writing developed for writing in stone has over time become detached from those original material conditions; because the imposing materiality of stone has diminished in an epoch characterized by extraction and upheaval; because the writing that is the object of this poem consists of a hastily scribbled chalk line on a wood fence—nevertheless literature only has recourse to a form of survival peculiar to stone. It falls back on what has fallen away. What is at stake for Benjamin in the lapidary style of Brecht's poem is not the matter of *survival* per se but rather a form of perseverance that obtains first and foremost for a mineral community. This claim turns on the observation that it is decidedly *not* the case that the words might "survive" the end of the world, as the translation of Benjamin's sentence in the *Selected Writings* gives, but rather that they might *outlast* or *endure* it: Benjamin's term is *überdauern ("outlast"),* not *überleben ("outlive").* In this way the poem about the loss of a life is also about the unsuitability of a purely biological concept of life (über*leben*) to inform the perseverance of messages across generations, if not also geological epochs. "Survival" must be understood as a paleonym. It is about the *survivance* of poetic words, to adopt a term articulated by Jacques Derrida and extensively adapted by Gerald Vizenor and other Indigenous scholars collected in *Survivance: Narrative of Native Presence*.[60] As a literary form of posterity, as Derrida writes, "in which the opposition of the living and the dead loses and must lose all pertinence, all its edge," survivance offers a form of survival "not thinkable on the basis of the opposition between life and death."[61] *Überdauern*, like survivance, consists in a spectral existence shared by bears, stones, fleas, mirrors, and professors—some of the protagonists of Vizenor's *Dead Voices*—as well as poems, novels, and "everything from which the tissue of living experience is woven."[62]

The subtle yet forceful shift from out*living* to out*lasting* the end of the world, from biological survival to bibliogeological survivance, establishes the basis for what Wai Chee Dimock more recently refers to as a planetary literature. Benjamin's commentary on a poem about a message that persists after the end not only of its author but conceivably also the world accords with Dimock's proposal to read literature as "an insult and an affront to the finiteness that is the norm of biological organisms and territorial jurisdictions."[63] For Dimock the temporally and spatially extended "life" of literary objects, exemplified in the case of Osip Mandelstam's reception of Dante, demonstrates how literature surpasses the scope of the nation and can come to encompass the entire planet. The durability of literature is due to its planetary form, which is forged through translation and the resulting availability of texts for a readership distributed across space and time. Although similarly employing a concept of life that would accommodate nonlife in his attempt to account for the vitality of translation in "The Task of the Translator"—"everything that has a history of its own, and is not merely the setting for history, is credited with life" (*allem demjenigen, wovon es Geschichte gibt und was nicht allein ihr Schauplatz ist, Leben zuerkannt wird*) (SW, 4:254–255; GS, 4.1:11)—Benjamin differs from Dimock in that he chalks up the durability of certain texts to their poetic form and their lapidary style, utilized not just to speak to future generations per se but in the first place to *outlast* the end of the world and to thereby anticipate the arrival of a radically unforeseeable reader and an equally unforeseeable reading.[64] But in at least one earlier image, a thought figure from *One-Way Street*, Benjamin points toward the possible reconciliation of the two vectors of planetary literature: "Commentary and translation stand in the same relation to the text as style and mimesis to nature: the same phenomenon considered from different aspect" (SW, 1:449; GS, 4:192).

Storied Matter: For a Multispecies War Primer

What, then, do these poetic words preserve and retain)? To what do these words hold fast, these words that no longer adhere to their background? They retain an echo, no matter how faint, of the classical past. By attributing an "inscription style" to these words, and by invoking the Horatian ode *Exegi monumentum aere perennius* ("I have raised a monument more permanent than bronze"), Benjamin evokes a continuum between modernity and antiquity and with it a greatly expanded set of coordinates for reading these lines. Yet these poetic words do not hold fast to the act of inscription, an *Inschrift*, as much as they do to the gesture of leaving a message, an *Aufschrift*, a word more appropriate for an adhesive label or graffiti.[65] *Exegious monument, aerily perennious*: H. C. E.'s rendering of the ode's opening line in *Finnegans Wake* ("our notional gullery is now completely complacent, an exegious monument, aerily perennious"), published in the same year that Benjamin wrote his commentary, conveys an antimonumentality similar to the antiauthoritarian poem that adopts a style reserved for monumental writing and which, in this act of citation, transforms ephemeral trace into indelible line.[66] Joyce's brilliant solecism—for *exiguous* (i.e., "negligible") read *exegi* ("I have raised"), for *aerily* (i.e., "insubstantially") read *aere* ("bronze")—reproduces the poem's (and Benjamin's) discovery that any recourse to the monumentality of antiquity is tenuous if not untenable. This contradiction is registered in and performed by Benjamin's commentary. The "character" of the *War Primer* poetry collection arises from the contradictions that Benjamin spells out—*in words which, through their poetic form, are expected to outlast the coming end of the world, the gesture of a message left on a wooden fence is retained, which someone persecuted hastily scrawls.* However, his earlier reference to the poem's character (from the Ancient Greek word meaning "an instrument used for engraving") is already rendered untenable, since the poem does not avail itself of an image of engraving or any other incisive mark. But the dialectical maxim announced at the outset of the commentary—*overcome difficulties by multiplying them*—suggests that *Inschrift* and *Aufschrift*, writing-in and writing-on, stand in a dialectical rather than oppositional relationship. A nondialectical, which is to say, a stratigraphic and Holocenic logic of superposition is suspended. This is tied up with the horizon of liberation to which the poem also holds fast. "The Invincible Inscription" ("Die unbesiegliche Inschrift"), another poem in the *Svendborg Poems*, testifies to liberatory possibility afforded by this dialectic of permanence and transience: a soldier imprisoned in Italy scratches *Hoch Lenin* ("Long Live Lenin") with a copying pencil on the wall of his cell, and, despite multiple attempts by the warden to have it painted over, the words remain visible, until a mason scrapes them off letter by letter, in the process turning this *Aufschrift* into an *Inschrift*, a superficial scribble into an indelible inscription ("Now remove the wall!" commands the soldier to the warden in the closing line.) As this poem also makes evident, both Brecht's poems and Benjamin's commentary are also shot through with fears of political persecution and arrest, another form of *festhalten*: intertwined with the recognition of the portability of poetic words is the authors' awareness of their own deportability.

But these chalked lines retain other traces, traces that can be found in what Benjamin might term the poem's *Sachgehalt* or "material content" and which is the proper domain of commentary, traces which extend the context of the poem back to another extinction event some 66 million years ago. In line with Benjamin, Clark, and Dimock's injunctions to consider the multiple scales in which literary texts are embedded, a commentary today might observe how the material and medium of chalk (*Kreide*) brings into play a set of spatial and temporal coordinates at a geologic scale: namely, those of the Cretaceous Period or *Kreidezeit* characterized by the deposition of immense chalk beds in what is today Northwestern Europe and lasting from 145 to 65 million years ago, bounded by the Cretaceous/Paleogene mass extinction event. Those moribund lines, written in chalk by someone who has already fallen, register a former extinction event as a condition of their possibility, while also foretelling, as an index of a species' immoderate extraction of mineral resources, a future one. In this way they anticipate the stratigraphic question of the Anthropocene—the problem of how to read the legacy of an extinct species—that enthralls us today.

The didacticism of chalk lies not only in the words one writes with it but also in its capacity as what material ecocriticsm would call "storied matter," that is, its capacity to be read as "a corporeal palimpsest in which stories are inscribed."[67] As a biogenic rock, one whose immense deposits serve as a medium to convey the planet's history, chalk functions as both stylus for didactic inscription and substrate of nonhuman inscription. Its corporeality encompasses both Iovino and Oppermann's definition of storied matter and MacKay's definition of corporeality as a "mediated material substrate" on which a text in the inscriptive style depends. The history of reading chalk as storied matter itself has a storied history, extending at least as far back as biologist Thomas Henry Huxley's oft-cited 1868 lecture "On a Piece of Chalk," delivered to the working men of Norwich during a meeting of the British Association for the Advancement of Science. In it Huxley describes how a "great chapter of the history of the world is written in the chalk" and proposes methods "of making the chalk tell us its own history" while inviting his audience "to spell that story out together."[68] It is a story of an everyday object bearing witness to the depths of time that vastly eclipse human history but through which that history is mediated.

Like the multiple extinction events unwittingly inscribed in Brecht's poem, the piece of chalk that is the subject of Huxley's address—a stylus composed of the material remains of extinct organisms that accumulated and petrified under the crushing weight of the oceans—appeals to a mineral community.

For Huxley "few chapters of human history have a more profound significance for ourselves" than the chapter written in chalk: "the man who should know the true history of the bit of chalk which every carpenter carries about in his breeches-pocket" will have a "truer conception of this wonderful universe, and of man's relation to it, than the most learned student who is deep-read in the records of humanity and ignorant of those of Nature."[69] This gesture toward a common mineral ground also indicates the common moribundity expressed in the chalked line in Brecht's poem. It bears recalling, with Huxley, that chalk consists of the skeletal remains of coccolithophores, an algal single-celled organism that belongs (by most accounts) to the Kingdom Protista, and in this connection it also bears mentioning in passing that these feature in the final section of Freud's *Beyond the Pleasure Principle* as evidence of a universal death drive, a drive to return to an inorganic state, operating in the simplest as well as the most complex forms of life. Setting aside the anthropomorphisms, tautologies, and contradictions that animate Freud's discussions of protists, his attempt to articulate a mineral community in which the animate and the inanimate are intimately linked—via the becoming-inorganic of the organism—might bear out in Brecht's scene of writing.[70] Although protist-bearing chalk is the material in which the words that will bear the coming catastrophe (*They Want War*) are written, chalk is itself a medium of extinction: in the line chalked on a wall we might read, as Kathryn Yusoff writes of the rising CO_2 levels resulting from fossil fuel extraction and consumption, "the trajectory of one extinction event feeding another."[71] In reanimating a fallen partisan's words as they were written in chalk, "The War Primer" also performs the reanimation of another extinction event that unfolded in deep time, and which, as a small monument to Homo sapiens' destratification of Cretaceous deposits, can be folded into the Anthropocene's generalized state of war. Instead of the extinction of a medium (chalk), the poem presents *extinction as media*, along the lines of John Durham Peters's definition: "containers of possibility that anchor our existence and make what we are doing possible." [72]

This admittedly deranged commentary on the Brecht poem, encouraged by a widespread desperation today, remains consistent with the emergent behavior of the literary text identified by Benjamin's commentary. The lapidary style of Brecht's poem remains forceful today because it both archives and animates what was thought to be lost to posterity. In this way the aspect of Brecht's poem most interesting to Benjamin also comprises the shock of the Anthropocene: the retention, resilience, remediation, and reanimation of what was not intended for posterity. *Exiguous monument, aerily perennious*: the enduring signature of the human, we are told, will not have been the inscriptions made in marble but more so the enduring, accidental monumentality of the nonbiodegradable artifacts that accumulate on the surface (and in the excavations) of the Earth as an affront to the finitude of all individual authors and even the species: infrastructural footprints, fossil records of novel extinction events, radioactive nucleotides, exploding CO_2 levels, and other signatures of war and extraction. As Zalasiewicz observes with respect to the coming ecological calamity: "The writing is already on the wall."[73] If the *War Primer* poems serve as a primer for reading the writing on the wall at the onset of World War II, they might again serve as a primer for reading strata at the onset of the general state of war known as the Anthropocene. One problem with this reading, however, is that for Brecht and Benjamin these trace fossils of revolt issue from anti-imperial and antioccupation agents, whereas in most contemporary accounts of the Anthropocene what endures attests to the resilience of, rather than the resistance to, imperialism, fossil capitalism, and racial violence. This contradiction remains to be resolved. I have endeavored to offer a few steps toward overcoming these contradictions: a reading of the Anthropocene, facilitated by Mark Bradford's *Help Us*, not as an act of mastery but as an attestation of a multispecies vulnerability; a reading of the Anthropocene, facilitated by Jeremy Davies's observation, in which a discrete human signature will have been illegible if not insignificant; a reading of the Anthropocene, facilitated by Brecht's late Holocene lapidary lyric, in which the indelible traces of extractivist societies will include the resistance to those societies' violence; and, finally and perhaps most fundamentally, a reading of the Anthropocene, with Benjamin's commentary, that insists on the geological as a matter of and for *reading* and thus insists on perpetually rereading polysemous inscriptions, whether epigraphic or stratigraphic, that as such resist, sooner or later, our anthropocentric modes of reading.

Acknowledgments

Gut Ding will Weile haben (Good things take a while), observes a miner in Goethe's *Wilhelm Meister's Journeyman Years*. This book is certainly a thing that has taken its time and the time of many others. I would like to start by recognizing their contributions, beginning with the support it received at Yale from the entire German Department, especially my advisor Rainer Nägele, as well as Henry Sussman and Carol Jacobs. Additional support during this stage came from the German Academic Exchange Service (DAAD) and the opportunity to participate in a graduate student colloquium at Humboldt University in Berlin on "The Knowledge of Literature," as well as the National Endowment for the Humanities (NEH) and the opportunity to participate in a summer seminar on Walter Benjamin's later writings at the University of California, Irvine. I thank Joseph Vogl for the invitation to the former and Alexander Gelley for the invitation to the latter, as well as participants of both for their feedback. Prior to all of these institutions and for the duration of a master's degree, the German Department at the Johns Hopkins University provided an intellectually intense setting where I first encountered most of the main reference points in this book and where (including the Ottobar) I started some of my longest-lasting friendships.

During successive stages, the book expanded thanks to invitations to present related work for a number of special occasions: for the Interdisciplinary Humanities Center at the University of California, Santa Barbara; for the "Environmental Humanities and German Studies Symposium" at the University of Toronto; for the "Symposium on Elements in Modern Thought and Literature" at New York University; for the symposium "Ecological Archives: Histories of Environment in German Studies" at Emory University; and for the Environments and Societies Colloquium at the University of California, Davis. My gratitude to the organizers of those events as well as the organizers of the German Studies Association's Environmental Studies Network, which has provided a number of further occasions to present work related to the book.

A memorable summer spent as an urban fellow at the Exploratorium Museum of Science, Art, and Human Perception provided space to write and an impetus and forum to engage multiple publics. I would like to thank the other fellows for their ongoing fellowship, Marina McDougall and the entire Center for Art and Inquiry for all of their support along the way, Susan Schwartzenberg for her role in developing and maintaining the most remarkable forum in the Bay Area for conversations on landscape and the environment, and the SEED fund for supporting the urban fellow program. Many others have offered less formal but no less meaningful forms of fellowship along the way: Ansgar Mohnkern, Josh Alvizu, Paul North, Tove Holmes, Ellwood Wiggins, Matthew Au, Timothy Attanucci, Michael Powers, Bernhard Malkmus, Volker Langbehn, Amir Eshel, Scott Lankford, John Gillis, Tina Gillis, Helen Mehoudar, Samuel Mehoudar, Ilana Halperin, Caroline Schaumann, Heather Sullivan, Jamie Kruse, Elizabeth Ellsworth, Rich Watts, and Michael Swaine.

I am especially grateful for my wonderful colleagues in the Department of Germanics at the University of Washington, in particular Sabine Wilke for her mentorship, Brigitte Prutti for encouraging my completion of the book during her tenure as chair, and Michael Neininger and Stephanie Welch for all of their support. Much gratitude is owed to the Walter Simpson Center of the Humanities at the University of Washington and director Kathleen Woodward for the lively intellectual exchange and teaching relief provided by the Society of Scholars fellowship as well as the conversations generated by the Crossdisciplinary Cluster on the Anthropocene that I was able to co-organize for the past three years with my brilliant colleague Jesse Oak Taylor.

For a book this fascinated by erratics, it will come as no surprise that it went through many iterations in many different places. In addition to the readers mentioned above, individual chapters have benefited from the critical attention given to them by Eric Downing and Tobias Menely. Special thanks also to the editors of earlier versions of chapter 1, which was published in *Readings in the Anthropocene: The Environmental Humanities, German Studies, and Beyond*, edited by Sabine Wilke and Japhet Johnstone (New York: Bloomsbury, 2017) ; and chapter 2, which appeared in *Goethe Yearbook* 22, no.1 (2015). I also want to thank Sabine Wilke, once again, for inviting me to edit and introduce "Literature und Geology," a special issue of *Literatur für Leser* 39, no. 1 (2016) that inspired several of the ideas and concepts that guide the introduction. The greatest editorial service of all has been provided by Fordham University Press, in particular acquisitions editor Tom Lay, managing editor Eric Newman, and copy editor Michael Koch for their guidance and their indispensable support for this book. I also would like to thank the anonymous reviewers, whose excellent insights informed my final revisions.

I am extremely grateful to Michael Levine for his ongoing and unwavering encouragement in the form of the insightful commentary on multiple pieces of this book and the multifaceted mentoring that extends from a sun-drenched Irvine to a snow-covered New Brunswick as well as many stations on I-95 in between. Equally profound is my gratitude to Henry Sussman for his conviviality first of all, for being such an inspiring co-conspirator on our blog *Feedback*, and also for that lift to Albany for the Institute of Critical Climate Change conference that was so transformative for my work.

My parents have supported my love of reading from the very beginning, and I could not imagine this book without them. If they inadvertently precipitated my fascination with German literature and philosophy by moving to Carlsbad (California) when I was very young, my father's bewildering gift of Kant's *Critique of Pure Reason* and Schopenhauer's *World as Will and Representation* for my sixteenth birthday (largely so he could have someone to read these with) turned out to be a more direct impetus. For a long time his library had been slowly absorbed into mine, and only more recently have we reached a bibliographic watershed where the flow has been in the other direction. My mother first taught me about actual watersheds and so many other ecological relationships and thereby engendered the receptivity to these relations that underlies this entire book. I also thank my parents-in-law for all of their support during the research and writing of this book, as well as my aunt-in-law, the most avid reader and writer in that family, for our many conversations on all things literary.

Sophie Lebrecht was there for me during every stage of this book. I cannot begin to convey my gratitude to her, but her vigorous advocacy coupled with her fierce criticism deserves my deepest appreciation and admiration. Theo and Charlotte joined us during a more advanced stage and have illuminated everything ever since.

Notes

Introduction

1. McKibben, *Eaarth*, 2–3.
2. McKibben, 2.
3. Most recently and forcefully, Yusoff, *Billion Black Anthropocenes*.
4. Goethe, *Journeyman Years*, 113.
5. Celan, *Der Meridian*, 5. I also translate *Neigungswinkel* as "unique bent" following Tobias, *Discourse of Nature*, 115.
6. Piper, "Mapping Vision," 33.
7. Braungart, "Poetics of Nature," 29.
8. Buckland. *Novel Science*, 13.
9. See Ellsworth and Kruse, "Introduction," 6–12.
10. Caruth, *Literature*, 92. The traumas associated with the Anthropocene might be one such instance. Recently critical geographers Zoe Todd, Heather Davis, Simon Lewis, and Mark Maslin have forcefully argued that the trauma of settler colonial genocide in the Americas is registered geologically, and accordingly the commencement of a novel geologic epoch currently identified as the Anthropocene should be pushed back several centuries to 1610 and the start of a colonial period with the so-called Columbian Exchange. Even though I explore other framings of the Anthropocene, I do not want to lose sight of the one that would begin in 1610 and that would, in the words of Zoe Todd and Heather Davis, "understand the current state of ecological crisis as inherently invested in a specific ideology defined by proto-capitalist logics based on extraction and accumulation through dispossession." Davis and Todd, "On the Importance of a Date," 764.
11. Schwab, *Haunting Legacies*, 7.
12. Bonneuil and Fressoz, *Shock of the Anthropocene*, 199.

13. See Freud, "Difficulty in the Path of Psychoanalysis," 22; Freud, "One of the Difficulties of Psycho-Analysis," 347–356.

14. Gould, *Time's Arrow*, 2: "Freud omitted one of the greatest steps from his list, the bridge between spatial limitation of human dominion (the Galilean revolution), and our physical union with all 'lower' creatures (the Darwinian revolution). He neglected the great temporal limitation imposed by geology upon human importance—the discovery of 'deep time' (in John McPhee's beautifully apt phrase)." The "discovery" of deep time may be far more allochronic than is often assumed; see Prete, "Being the World Eternal," 292–317.

15. Morton, *Hyperobjects*, 18.

16. Boehme, "Das Steinerne," 119–141.

17. Sullivan, "Organic and Inorganic Bodies," 37.

18. Povinelli, *Geontologies*, 14.

19. Ransom, "Psychologist Looks at Poetry," 582.

20. Adorno, *Aesthetic Theory*, 322.

21. Celan, "Ansprache," in *Gesammelte Werke*, 3:185.

22. Parikka, *Geology of Media*, 20.

23. Cohen, "Anecographics," 42.

24. Heise, *Sense of Place*, 67.

25. Cohen, "Anecographics," 44.

26. In *Loiterature* Ross Chambers refers to a genre of modern literature marked by a counter-disciplinary resistance to the prevailing structuring in space and time of work. Through primarily referring to flâneur realism, "loiterature" could be extended to encompass those dilatory narratives—whose protagonists are primarily peripatetics—that formally engage the deep time of the Earth, from Adalbert Stifter's *Indian Summer* to Peter Handke's *Slow Homecoming*. See Chambers, *Loiterature*; Lacan, "Lituraterre"; and Schestag, "Interpolationen," 55.

27. Moraru, *Reading for the Planet*, 94–95.

28. Kuh, *Zwei Dichter*, 468: "kleinliche Detailmalerei unwesentlicher Dinge."

29. Cohen, *Stone*, 8.

30. Blanchot, "Two Versions of the Imaginary," 254–263.

31. Dawson, *Soldier Heroes*, 1994, 48.

32. Caillois, *Œuvres*, 1155: "L'imagination n'est rien de plus qu'un prolongement de la matière."

33. Cohen and Colebrook, "Preface," 9.

1. Of Other Petrofictions: Reimagining the Mine in German Romanticism

1. Ghosh, "Petrofiction," 29–34.

2. Sullivan, "Material Ecocriticism and the Petro-Text," 417.

3. Ziolkowski, *German Romanticism*, 25. For an account of British geology in this period see Noah Heringman, *Romantic Rocks, Aesthetic Geology* (Ithaca, NY: Cornell University Press, 2004).

4. Parikka, *Anthrobscene*, 6.
5. Ziolkowski, *German Romanticism*, 21.
6. Yusoff, "Geologic Life," 780.
7. Uglietti et al. "Widespread Pollution," 2349.
8. Moore, *Capitalism in the Web of Life*, 185.
9. "An manchen Orten waren die Gängen unten und an den Wänden etwas naß; auch sahen wir ein Paar Pumpen, das Wasser heraufzuschaffen, und eine große Art von Schacht, Radstube genannt, für eine neue Maschine zu dieser Absicht." Wackenroder, *Werke*, 167. All translations of Wackenroder are mine.
10. Crutzen, "Geology of Mankind," 23.
11. See Myles Jackson, "Natural and Artificial Budgets," 427–428. The sketch shown in figure 1 is likely from that trip.
12. Lovelock, *Rough Ride*, 18.
13. Novalis, *Heinrich von Ofterdingen*, in *Schriften*, 1:253; my translation.
14. Clark, *Inhuman Nature*, 81–106; Bennett, *Vibrant Matter*, 7–8; Parikka, *Geology of Media*, 32; Ford, "Romanthropocene."
15. Sullivan, "Dirty Nature," 121.
16. Rigby, "Mines Aren't Really Like That," 116.
17. Morton, *Hyperobjects*, 109.
18. Rigby, *Topographies of the Sacred*, 149.
19. Lyon, "Disorientation in Novalis," 85–103.
20. Tieck, "Der Runenberg," 186, 191; my translation. Hereafter cited in text as R followed by page number in parentheses.
21. Rigby, "Mines Aren't Really Like That," 117.
22. Ziolkowski, *German Romanticism*, 28.
23. Allen, "Mineral Virtue," 128.
24. Parikka, *Anthrobscene*, 6.
25. Mumford, *Technics and Civilization*, 70.
26. See Safranski, *Romanticism*, 62.
27. Wackenroder, *Werke*, 167.
28. Wackenroder, 166.
29. Wackenroder, 167.
30. Wackenroder, 167.
31. Grosz, *Becoming Undone*, 20.
32. Wackenroder, *Werke*, 166.
33. Grosz, *Becoming Undone*, 11.
34. Cohen, *Stone*, 10.
35. Cited in Ziolkowski, *German Romanticism*, 32.
36. Ziolkowski, 29.
37. Schubert, *Ansichten*, 200–201; my translation.
38. Translation by Sullivan, "Ruins and the Construction of Time," 28. The original reads: "Die Thränen stürzten mir aus den Augen; es war mir, als wenn das Innerste der Erde seine geheimnißvollste Werkstatt mir eröffnet hätte; als wäre die

fruchtbare Erde, mit ihren Blumen und Wäldern, eine zwar anmuthige, aber leichte Decke, die unergründliche Schätze verbarg, als wäre sie hier zurückgezogen, abgestreift, um mich in die wunderbare Tiefe hinabzuziehen, die sich eröffnete. Der Eindruck war ein durchaus fantastischer, und es mag eine lebhafte Darstellung von diesem Eindrucke gewesen fein, welche Tieck veranlaßte, seine Novelle, den Runenberg, zu schreiben . . . Tieck hat gestanden, bei dieser Novelle an mich gedacht zu haben." Steffens, *Was ich erlebte*, 22–23.

39. Sullivan, "Organic and Inorganic Bodies," 33–34.
40. See Povinelli, *Geontologies*, 130.
41. Kuzniar, "Stones that Stare," 54.
42. Kuzniar, 55.
43. Bennett, *Vibrant Matter*, 56.
44. Deleuze, *Pure Immanence*, 29.
45. Bennett, *Vibrant Matter*, 55.
46. Bennett, 60.
47. Clark, *Inhuman Nature*, xiv.
48. For a recent philosophical account of the eccentricity of the Earth, see Neyrat, *Unconstructable Earth*.
49. Campe, "Rauschen of the Waves," 23.
50. See Gasperi, "On the Language of Nature," 405–430.
51. In German: "ein dumpfes Winseln im Boden, das sich unterirdisch in klagenden Tönen fortzog, und erst in der Ferne wehmütig verscholl."
52. Parikka, *Geology of Media*, 4.
53. I am indebted to Heather Sullivan for providing a wonderful translation of these lines in "Organic and Inorganic Bodies" (28), which I have modified here.
54. Merola, "Mediating Planetary Attachments," 253.
55. Sullivan, "Organic and Inorganic Bodies," 36.
56. Kuzniar, "Stones that Stare," 51.
57. Ziolkowski, *German Romanticism*, 51.
58. Benjamin, *Selected Writings*, 1:145. Benjamin, *Gesammelte Schriften*, 1.3:55: "die Bedingtheit jeder Objekterkenntnis in einer Selbsterkenntnis des Objekts."
59. Benjamin, *Selected Writings*, 1:145. Benjamin, *Gesammelte Schriften*, 1.1:55: "In allen Prädikaten, in denen wir das Fossil sehen, sieht es uns."
60. Novalis, *Die Werke Friedrich von Hardenbergs*, 3:650.
61. Kuzniar, "Stones that Stare," 53.
62. See Laurence A. Rickels, "Mine," http://vv.arts.ucla.edu/terminals/rickles/rickles.html.
63. The German reads: "Es war ein Mann in einem ganz zerrissenen Rocke, barfüßig, sein Gesicht schwarzbraun von der Sonne verbrannt, von einem langen struppigen Bart noch mehr entstellt."
64. Nixon, "Anthropocene."
65. Ziolkowski, *German Romanticism*, 53.

2. Goethe's Erratics: Wandering through Deep Time

1. Goethe, *Sämtliche Werke*, 10:287. Hereafter cited in text as FA (*Frankfurter Ausgabe*) and part number (I for works, II for letters), *volume, and page number in parentheses*. For the English version, see Goethe, *Collected Works*, 10:113. Hereafter cited in text as CW *and volume and page number in parentheses*.
2. Frank, "Steinherz und Geldseele. Ein Symbol im Kontext," 253–387.
3. Morton, "Thinking Ecology," 265–293.
4. Bennet, *Vibrant Matter*. On a "geopoetics" within the German lyrical tradition, see Schellenberger-Diederich, *Geopoetik*. See also Deleuze and Guattari, "Geophilosophy"; Westphal, *Geocriticism*; Moraru, *Reading for the Planet*; Woods, "Cosmic Passions"; Caverero, *Inclinations*; and Harris, Turner, and Noeck, "Rock Record." In their 2013 film *Goodbye Gauley Mountain: An Ecosexual Love Story* (New York: Kino Lober), ecosexual pioneers Beth Stephens and Annie Sprinkle chronicle their marriage to the Appalachian Mountains in an attempt to bring awareness to the coal mining practice of mountain top removal. For an overview of the proposed marriage between queer theory and ecocriticism, see Morton, "Guest Column: Queer Ecology." Morton acknowledges the pioneering work of Mortimer-Sandilands and Erickson, *Queer Ecologies*. For a discussion of the Great Acceleration in the context of the Anthropocene, see Steffen, Crutzen, and McNeill, "Anthropocene."
5. Crutzen, "Geology of Mankind," 23.
6. Goethe, *Die Schriften zur Naturwissenschaft*, I.11:9. Hereafter cited in text as LA (short for Leopoldina Ausgabe) and *part number* (I for works, II for letters), volume, and page number in parentheses.
7. See Sullivan, "Collecting the Rocks of Time," 341–370; and Powers, "The Sublime," 35–56. See also Görner, "Granit: Zur Poesie eines Gesteins"; Muenzer, "Ihr ältesten, würdigsten Denkmäler der Zeit," 181–198; and von Engelhardt, *Goethe im Gespräch mit der Erde*, 83–118.
8. Piper, "Mapping Vision," 33.
9. Piper, 33.
10. This turn of phrase comes from Hutton, "Erratic Imaginaries," 111–124.
11. Wyder, "Goethes geologische Passionen," 136.
12. Wyder, "Gotthard, Gletscher und Gelehrte," 42.
13. See Thüsen, "Goethes Vulkane," 265–280.
14. Bisanz, "Birth of a Myth," 190.
15. Caverero, *Inclinations*, 11.
16. "Findlinge nennt man sie, weil von der Brust, / Der mütterlichen sie gerissen sind, / In fremde Wiege schlummernd unbewußt, / Die fremde Hand sie legt wie's Findelkind." Droste-Hülshoff, "Die Mergelgrube," 50; my translation.
17. Schellenberger-Diederich, *Geopoetik*, 208.
18. Prior to Jillian DeMair's recent essay and an earlier version of this chapter, the only piece of scholarship to consider seriously the poetological implications of the

geological erratic in German literature was Irmgard Wagner, "Der Findling: Erratic Signifiers in Kleist and Geology." See also DeMair, "Geological Uncertainty," 41–72. On the image of the human foundling in this period, see Batten, *Orphaned Imagination*.

19. For a more extensive overview of Goethe's engagement with the blocks, see Cameron, "Goethe," 751–754.
20. Wyder, "Gotthard, Gletscher und Gelehrte," 100.
21. See Engelhardt, "Did Goethe Discover the Ice Age?," 123–128.
22. Müller-Sievers, *Science of Literature*, 226.
23. Müller-Sievers, 226.
24. Goethe, *Goethes Gespräche*, 571.
25. Goethe, 226.
26. This, though, does not excuse its anti-Semitic outbursts. See Schutjer, "Beyond the Wandering Jew," 389–407.
27. See Krüger, *Discovering the Ice Ages*. On Goethe's contribution, see section 3.5, "Minister of Mining Goethe Has His Own Ideas," 109–129; Krüger concludes: "Goethe was one of the earliest and most independent advocates of ice-age theory" (129).
28. Trexler, *Anthropocene Fictions*, 23.
29. Massey, "Landscape as a Provocation," 34.
30. Morton, *Hyperobjects*, 196.
31. Steffen et. al., "The Anthropocene," 614.
32. Morton, *Hyperobjects*, 1.
33. Clark, *Inhuman Nature*, xi, xiv.
34. Bennett, *Vibrant Matter*, 9.
35. See, for example, the aphorism that inaugurates Makarie's Archive in the collection of brief sentences at the end of the novel's third and final book: *Die Geheimnisse der Lebenspfade darf und kann man nicht offenbaren; es gibt Steine des Anstoßes, über die ein jeder Wanderer stolpern muss; Der Poet aber deutet auf die Stelle hin* (FA, I.25:646). Relentlessly anapocalyptic, the passage represents an impasse in understanding posed by necessarily unanticipatable obstacles on a metaphoric path of life and can be counted among those "open secrets" (*offenbare Geheimnisse*) that figure so significantly in Goethe's oeuvre, most pronounced in the letter to Carl Jacob Ludwig Iken on September 27, 1827.
36. Clark, "Scale," 164.
37. Alley, *Two-Mile Time Machine*, 120.
38. Wagner, "Der Findling," 290.
39. Wagner, 294. *Irrblöcke* was, however, used by biologist Lorenz Oken in the 1830s, as Schellenbgerger-Diederich shows, and can even be found in a 2000 Lexicon of Geosciences. See Schellenberger-Diederich, *Geopoetik*, 210–211 as well as DeMair, "Geological Uncertainty," 48.
40. Bachelard, *Formation*, 24.
41. For a discussion of Werner's considerable influence in the geology and literature of the *Goethezeit*, see Haberkorn, *Naturhistoriker und Zeitenseher*.

42. Bachelard, *Formation*, 24. Yet in the identification of thinking with movement a conceptual sclerosis develops in his own thinking. The "obstacle," the word on which *The Formation of the Scientific Mind*, hinges, assumes the status of what Bachelard in another context calls the "verbal obstacle"—the fixation on a single word. Perhaps, the most sclerotic conceptualization of all consists in the identification of thinking with forward movement. "Where are the sources of the mind's movement to be found? How can scientific thought recover and find a way out of this situation [of conceptual sclerosis]?" Bachelard asks. To which Goethe might respond: *in the block itself.*

43. Bachelard, *Formation*, 25.
44. Grimm, *Deutsches Wörterbuch*, 5:col. 3884–3885.
45. See Reynard, "Protecting Stones," 4.
46. For a review of this dispute see David Roger Oldroyd, "The vulcanist-neptunist debate reconsidered," *Journal of Geological Education* 19 (1997): 450–454.
47. Wilke, "Toward an Environmental Aesthetics," 273.
48. Charpentier, *Essai sur les glaciers*, v. See also Cameron. "Goethe," 751–754.
49. Pörksen, "Raumzeit," 101–127.
50. Demair, "Geological Uncertainty," 41.
51. Glenn Collins, "The Very Cold Case of the Glacier," *New York Times*, September 14, 2005, B1.
52. Staiger, *Goethe*, 3:135.
53. Broch, "James Joyce," 206.
54. Goethe, *Goethes Gespräche*, 3.2:571. Martin Bez sees the aggregate form as a key figure for understanding the structure of the *Wanderjahre*; see Bez, *Goethes "Wilhelm Meisters Wanderjahre."* See also Shu Ching Ho, "Knochenbau der Erde als Konstruktion der Welt: zur Bedeutung der Geologie in 'Wilhelm Meisters Wanderjahren,'" *Goethe Jahrbuch* 125 (2008), 122–135.
55. Gideon, *Zur Darstellungsweise*, 24.
56. Piper, "Mapping Vision," 32.
57. Buell, *Environmental Imagination*, 84.
58. Latour, *Politics of Nature*, 237.
59. Engelhardt, *Goethes Weltansichten*, 244.
60. Engelhardt, 244. Engelhardt does not consider the possibility that Goethe was drawing on a much older topos, associated with Agricola's *De ortu et causis subterraeorum* of 1546, of a mineral kingdom *in statu nascendi*. See Adams, "Generation of Stones," 77–136.
61. Sullivan, "Faust's Mountains," 116.
62. Boyle, *Goethe*, 325.
63. Buell, *Environmental Imagination*, 133–179.
64. Piper, "Rethinking the Print Object," 137.
65. Goethe, *Goethe's Werke*, 2.10:92–94.
66. This dismissal of any attempt to reinstate these displaced bodies to a distant home will also crop up in the words of Mephistopheles as he confronts the geological problem of these foreign rock bodies with Faust: "The heavy chunks lie where they

don't belong, / and what ballistic force can be the explanation? / Philosophers and scientists are at a loss: / there is the rock, they say, you'll have to let it lie, / since we are hopelessly confounded by it." (FA, I.7:393; CW, 2:255).

67. Nixon, *Slow Violence*, 13.

68. For a thorough account of climate change fiction in Germany, see Goodbody, "Melting Ice," 92–102.

69. Chakrabarty, "Climate of History," 213.

3. Many Stranded Stones: Stifter's Spectral Landscapes

1. Webb, "Tedium, Tragedy, and Tar."
2. Sebald, "Bis an den Rand der Natur," 22.
3. Kuh, *Zwei Dichter Österreichs*, 468.
4. Koschorke, *Geschichte des Horizonts*, 280.
5. Frederick, *Narratives Unsettled*, 133. The idiom "indian summer" is increasingly suspect of having derogatory connotations. It is an unfortunate translation of the German title *Nachsommer*, which refers to abnormal seasonality and disrupted periodicity without resorting to a potentially derogatory term. For a fuller discussion of the English-language translation, see Latimer, "On Translating Stifter's 'Nachsommer.'" For a thoughtful and urgent consideration of German and Indigenous Studies, see Watchman et al., "Building Transdisciplinary Relationships: Indigenous and German Studies." Hereafter, Wendell Frye's English translation is cited in text as *IS* with page number in parentheses.
6. Strowick, "'Dumpfe Dauer,'" 187.
7. Schnyder, "Schrift—Bild—Sammlung—Karte," 235–248.
8. See Braungart, "Der Hauslehrer," 110.
9. Coen, *Climate in Motion*, 67.
10. Coen, 2
11. Coen, 142.
12. Stifter, *Werke und Briefe*, 2.2:11. Hereafter cited in text as *HKG* and volume and page number in parentheses. With the exception of *Indian Summer* all English translations are mine unless otherwise noted.
13. Lukács, "Erzählen oder Beschreiben?," 218.
14. Swales, *German Novelle*, 28.
15. Vischer, "Aesthetik oder Wissenschaft des Schönen," 364.
16. See Attanucci, *Stories from Earth*; and Holmes, "An Archive of the Earth."
17. Schnyder, "Die Dynamisierung des Statischen," 553.
18. Sebald, "Bis an den Rand der Natur," 18.
19. Bennett, *Vibrant Matter*, vii.
20. Benjamin, *Gesammelte Schriften*, 2.2:609.
21. Benjamin, 2.2:548–549; Benjamin, *Selected Writings*, 4:224.
22. Stifter, *Werke*, 7.
23. See Batten, *Orphaned Imagination*; and Geulen, "Stifters Sonderlinge."
24. Blackall, *Adalbert Stifter*, 12.

25. For a discussion of the problem of representation of motion and how in Stifter "motion is inscribed into the object itself," see Strowick, "Poetological-Technical Operations," 288.

26. Möseneder, "Stimmung und Erdleben," 32. See also Lindsey J. Brandt, "Tangled up in Truths: German Literary Conceptions of Nature between Romantic Science and Objective Empiricism," PhD diss., University of North Carolina at Chapel Hill, 2015.

27. Gr refers to Jeffrey L. Sammons's translation of Granite in German Novellas, 1:7–34.

28. Downing, Double Exposures, 45.

29. Santner, Stranded Objects, xiii.

30. Many-Colored Stones also references an anthological tradition and the act of gathering inhering in the act of anthologizing; the use of stones and minerals for the titles of stories was not at all uncommon in anthologies of the time, and a literary journal in Reichenberg even ran from 1848 to 1860 under the title Bunte Steine, gesammelt auf dem Gebiete des Wissenswerthen, Nützlichen und Angenehmen. Yet the ordering of the stories is also geologically informed, and it can be shown that a geological system furnishes their order—granite (Granit) is an igneous rock, calcite (Kalkstein) a sedimentary rock, and all others are of metamorphic origin—rather than any element immanent to the stories.

31. My remarks on Limestone are greatly informed by Lindsey K. Brandt's chapter on Stifter in her remarkable dissertation, "Tangled Up in Truths." I thank Eric Downing for bringing it to my attention.

32. Handke, Die Lehre. 93, 85; Handke, Slow Homecoming, 200, 195. For a much more extensive consideration of Handke's patchwork method, see Haubenreich, "Poetry, Painting, Patchwork."

33. Handke, Die Lehre, 78, 190; Slow Homecoming, 91, 199.

34. Sebald, "Bis an den Rand der Natur," 16.

35. An intensive and insightful comparison of the two is Prutti, "Zwischen Ansteckung und Auslöschung." For a more extensive study, see Wild, Wiederholung und Variation.

36. On this topic, see Sng, "Not Forgotten."

37. See Downing, Double Exposures, 44: "As we will see, he [the narrator of The Mountain Forest] will continue to tracer these two uneasily contradictory movements—one circular, turned back, stable, and closed, the other linear, descending, labile, and open-ended—throughout the narrative."

38. Swales and Swales, Adalbert Stifter, 144.

39. Compare with Die Pechbrenner: "Sie trug mich . . . in das Vorhaus hinaus. Dort . . . ließ mich auf dem Steine des Vorhauses, der zum Klopfen des Garnes dastand, weinend und schluchzen sitzen. . . . Ich konnte vor Schmerz über diese Wendung der Dinge nicht von dem Steine weg, es ließ mir kaum das Schluchzen herausstoßen, und das Herz war mir mit Schnüren zugezogen" (HKG, 2.1:14).

40. Freud, Die Traumdeutung, 127. Freud, The Interpretation of Dreams, 132.

41. Adorno, "Über epische Naivetät," 37.

42. Rosei, "Versuch über Stifter" (165): "Stifter hat sehr wohl gewußt oder zumindest gespürt, wie dünn das logische, sprachliche Parkett war, auf dem er sich bewegte." The question of the *Parkett* (from French *parquet*, "little park") already extends beyond any domestic situation (like the German *Flur*, "field"), particularly that of the notoriously unblemished floorboards of the *Rosenhaus* in *Indian Summer*. One of Stifter's suggestions for the collection of stories that became *Many-Colored Stones* was *Flursteine* (*Field stones*).

43. See Sabine Schneider, "Kulturerosionen."

44. "Wenigstens waren die Tuffplatten, die ihm als Unterlage dienten, schon mit einem dicken Antritte von Erde belegt, und da, wo sie unter die Dachtraufen hinaus ragten und rein gewaschen waren, mit sehr tiefen Löchern versehen" (*HKG*, 2.1:19).

45. See Micheler, "Vulkanische Idyllen."

46. Swales and Swales, *Adalbert Stifter*, 184.

47. "It is worth pointing out that the word *Erschütterung* ["shock"] contains the word *schütter* ["thin," "sparse"]. Wherever something collapses, rifts and gaps appear. As analysis has shown, the poem contains numerous passages in which words combine in a loose, unstable way to form the meaning. This contributes to its shocking [*erschütternd*] effect (Benjamin, *Selected Writings*, 4:224). "Man tut gut, sich darauf zu besinnen, daß das Deutsche in seinem Wort *Erschütterung* das Wort *schütter* stecken hat. Wo etwas zusammenstürzt, da entstehen Brüche und Leerstellen. Wie sich aus der Analyse ergibt hat das Gedicht [Bertolt Brechts "Gegen Verführung"] zahlreiche Stellen, an denen die Worte labil und locker zum Sinn zusammentreten. Das leistet seiner erschütternden Wirkung Vorschub" (Benjamin, *Gesammelte Schriften*, 2:548–549).

48. For two recent attempts to situate Stifter's work within the context of the Anthropocene, see Phillips, "Adalbert Stifter's Alternative Anthropocene"; and Ireton, "Adalbert Stifter and the Gentle Anthropocene."

49. Ireton, "Between Dirty and Disruptive Nature," 161.

50. The problem is also treated at length in Ketelsen, "Geschichtliches Bewußtsein"; and Wagner, "Schick, Schichten, Geschichte," 22.

51. See Sorg, *Gebrochene Teleologie*. The lack of narrative closure that Sorg stresses is another way of getting at Stifter's loose ends.

52. See Wenzel, "Stratigraphy and Empire."

53. Twellman, "Bleibende Stelle," 228.

54. Ketelsen, "Geschichtliches Bewußtsein" (310): "Er bewahrt in sich seit undenklichen Zeiten die Spuren der Generationen auf."

55. "If you were not so young, you would have seen the column that is no longer in existence but stood on the marketplace of Oberplan and on which you could read when the plague came and when it ceased [wenn du nicht so jung wärest, so würdest du auch die Säule noch gesehen haben, die jetzt nicht mehr vorhanden ist, die auf dem Marktplatze von Oberplan gestanden war, und auf welcher man lesen konnte, wann die Pest gekommen ist, und wann sie aufgehört hat]" (*HKG*, 2.2:44).

56. Lacan, "Lituraterre," 329. Frederick Leverett glosses the Latin *litura* as "the drawing or smearing of the wax of a writing-tablet over a letter or word, in order to efface it; the rubbing out of a letter or word; a rasure." Leverett, *A New and Copious Lexicon of the Latin Language*, 498.
57. Geulen, *Worthörig wider Willen*, 155.
58. Holmes, "Archive of the Earth," 291.

4. The Shock of the Earth: Benjamin's Unarticulated Ground

1. Ghosh, *Great Derangement*.
2. Cohen, "Climate Change," 27.
3. Nixon, *Slow Violence*.
4. Clark, "Blanchot," 20.
5. Latour, *Facing Gaia*, 116.
6. Bonneuil and Fressoz, *Shock of the Anthropocene*, 20.
7. Adorno, *Aesthetic Theory*, 244; translation modified.
8. For an analysis of the "gentle law," see Downing, "Real and Recurrent Problems: Stifter's Preface to *Many-colored Stones (Bunte Steine)*," in *Double Exposures*, 24–40.
9. Benjamin, *Selected Writings*, 1:112. Hereafter abbreviated in text as SW and volume number and page number in parentheses. The German reads: "Die Fähigkeit, irgendwie 'Erschütterung' darzustellen, deren Ausdruck der Mensch primär in der Sprache sucht, fehlt ihm absolut." Benjamin, *Gesammelte Schriften*, 2.1:209. Hereafter abbreviated in text as GS and volume number and page number in parentheses.
10. Selz, "An Experiment by Walter Benjamin," 147.
11. Woodard, *On an Ungrounded Earth*, 2.
12. Benjamin, *Correspondence*, 340; Benjamin, *Briefe*, 1:478. Unless otherwise noted, all translations from *Briefe* are mine.
13. Benjamin, *On Hashish*, 49. The phrase "*unregulierten Erdboden*" appears in both the original hash protocol of September 29, 1928—a text included in translation in the collection *On Hashish* but not in the *Selected Writings*—and the short story "Myslowitz-Braunschweig-Marseilles" ("Myslovice-Braunschweig-Marseilles"), which was published in the journal *Uhu* in 1930 and which is translated in the selected writings (SW, 2.1:386–393). In the translated protocol of protocol of September 29, 1928, "*unregulierten Erdboden*" appears as "irregular earth" whereas in "Myslovice-Braunschweig-Marseilles" it appears as "uneven ground." Relatedly, Benjamin writes of an "*unartikulierten Erdboden*" in the article "Haschisch in Marseille" ("Hashish in Marseilles"), published in December 1932, and which is translated in the *Selected Writings* as "unarticulated earth." Future citations refer to "Hashish in Marseilles" of December 1932. My thanks to Michael Powers for drawing my attention to the hash protocol of September 29, 1928.
14. Glissant, *Philosophie de la Relation*, 45. Cited in Wiedorn, *Think Like an Archipelago*, 113.

15. Benjamin, *Correspondence*, 393; *Briefe*, 1:552.
16. Benjamin, 393; 1:552.
17. Benjamin, 393; 1:552.
18. Provisionally it would seem that the product of the Ibizan notations from the Spring 1932 letter to Gretel Adorno consists in the nine short pieces falling under the title "Ibizan Suite" (*Ibizenkische Folge*) first published in the *Frankfurter Zeitung* on June 4, 1932, and later collected by Adorno in the *Denkbilder* collection. Yet the resumption of writing thought images should not have been all that surprising, considering that Benjamin, in a letter from October 1928, already mentions several supplements to *One-Way Street* (*Briefe*, 3:484). Likewise, in his introduction to the 1955 edition of Benjamin's edited writings, Adorno refers to plans for a second, expanded edition of *One-Way Street*, while Benjamin's posthumous papers include a supplemental list for *One-Way Street* (*Nachtragsliste zur Einbahnstraße*), which contains most of the titles from the *Ibizan Suite*, including "Downhill" (*GS*, 4.2:911). Moreover, the dates attributed to the twenty-eight pieces collected in *Denkbilder* show a more or less unceasing production of short pieces in the period following the publication of *One-Way Street* through his stay in Ibiza.
19. Eiland and Jennings, *Walter Benjamin: A Critical Life*, 214.
20. Mittelmeier, *Adorno in Neapel*.
21. Richter, *Thought-Images*, 45.
22. Proust, *Sodom and Gomorrah*, 210.
23. Grimm and Grimm, *Deutsches Wörterbuch*, 15, col. 1613: "zerbröckeltes, zerriebenes gestein, einzelne auf ebenem felde vorkommende blöcke, findlinge, verwandt mit schutt, schütten."
24. See Richter, "Toward a Politics of the Unusable."
25. Menely, "The Present," 489–490.
26. Menely, 490
27. Menely, "Anthropocene Air."
28. Levine, *Weak Messianic Power*, 16.

Epilogue: Dilapidated

1. Fogle, *Life on Mars*, 76.
2. Economically driven discussions of rare earth mineral scarcity increasingly (and disingenuously) adapt a rhetoric of endangerment and extinction. For a careful reflection on mineral finitude, see sci-fi writer Silverberg, "Reflections," 8–10; see also Povinelli, *Geontologies*, 30–56.
3. See Mirzoeff, "It's not the Anthropocene."
4. Hartman, "Inscriptions and Romantic Nature Poetry," 35.
5. Bann, "Inscription in the Garden."
6. Cohen, *Stone*, 145–146.
7. Lyell, *Principles of Geology*, 459.
8. Waters et al., "Anthropocene," 137.
9. Menely and Taylor, introduction to *Anthropocene Reading*, 6.

10. Boes, "Reading the Book of the World," 106.
11. Boes, 108.
12. Zalasiewicz, *Earth After Us*, 235.
13. Derrida, *Of Grammatology*, 19.
14. Szerszynski, "End," 179.
15. Szerszynski, 179.
16. Campe, "Form and Life," 54.
17. Frederick, *Narratives Unsettled*, 174.
18. Davies, *Birth*, 162.
19. Frederick, *Narratives Unsettled*, 174.
20. Zalasiewicz, *Earth After Us*, 24.
21. Hartman, "Inscriptions and Romantic Nature Poetry," 35.
22. Zalasiewicz, *Earth After Us*, 118.
23. MacKay, *Inscription and Modernity*, 2.
24. Zalasiewicz, *Earth After Us*, 6.
25. Hartman, "Inscriptions and Romantic Nature Poetry," 33, quoted in MacKay, *Inscription and Modernity*, 238.
26. Uglietti et al., "Widespread Pollution," 2349.
27. MacKay, *Inscription and Modernity*, 6.
28. MacKay, 5; Boes, "Reading the Book of the World," 97.
29. MacKay, *Inscription and Modernity*, 30.
30. MacKay, 18.
31. Zalasiewicz, *Earth After Us*, 239.
32. Zalasiewicz, 1.
33. Wenzel, "Stratigraphy," 179.
34. Parrika, "Geology of Media."
35. MacKay, *Inscription and Modernity*, 23. Paul Fry likewise reads the crucial "what" in Wordsworth's line from *The Recluse* ("words/Which speak of nothing more than what we are") as evidence of "the revelation of being itself in the nonhumanity that 'we' share with the nonhuman universe." See Fry, *Wordsworth*, x.
36. MacKay, *Inscription and Modernity*, 16.
37. Davies, *Birth*, 81.
38. Davies, 73.
39. Davies, 81.
40. Davies, 108–109.
41. Davies, 83.
42. MacKay, *Inscription and Modernity*, 16.
43. Joy, "Notes" [online lecture].
44. Boes and Marshall, "Introduction," 64.
45. See Fiorato, "Zum Paradigma."
46. Clark, "Scale," 156.
47. Clark, 139.
48. Clark, 164.
49. Clark, *Ecocriticism*, 62.

50. Clark, 65.

51. Benjamin, *Selected Writings*, 4:215. Hereafter abbreviated in text as SW and volume number and page number in parentheses. For the German, see Benjamin, *Gesammelte Schriften*, 2.2:540. Hereafter abbreviated in text as GS and volume number and page number in parentheses

52. This constellation of commentary and reduction will in fact reemerge, in another setting, shortly after the completion of the Brecht commentary. "We will have to return to the topic of reduction as an artifice of phatasmagoria," Benjamin writes to Gretel Adorno in 1940. "This passage [from Adorno's *Versuch über Wagner*] reminded me of one of my oldest projects. You may remember having heard me speak of it: I am referring to the commentary on Goethe's *Neue Melusine*." Benjamin, *Correspondence*, 627; *Briefe*, 1:844.

53. Morton, *Hyperobjects*, 160–161.

54. Szerszynski, "End," 174.

55. Wenzel, "Stratigraphy," 180.

56. See Clark and Hird, "Deep Shit."

57. Wagenknecht, "Marxistische-Epigrammatik," 546.

58. Lessing, "Zerstreute Anmerkungen," in *Lessing's Werke*, 14:169; Lessing, *Fables*, 169.

59. Clark, *Ecocriticism*, 62.

60. See Derrida, *Beast*; Vizenor, *Survivance*.

61. Derrida, *Beast*, 130/193.

62. Derrida, 132/195.

63. Dimock, "Literature for the Planet," 182.

64. It also bears noting in this regard that the disjunction between an anonymous partisan's graffiti and his posthumous reader also anticipates Benjamin's struggle to find a publisher for this commentary, whose typescript (along with a German version of the *Artwork* essay, *The Storyteller*, and the convolutes later translated and published as the *Arcades Project*) was deposited in 1940 in the Bibliothèque nationale as Benjamin fled Paris.

65. For a thoughtful reading of the poem in the context of the relationship between language and architecture, see Miezskowski, "Writing Is on the Wall."

66. Joyce, *Finnegans Wake*, 57.

67. See Iovino and Oppermann, "Stories," 451.

68. Henry Huxley, "On a Piece of Chalk," 156.

69. Huxley, 156.

70. See Roof, "From Protista to DNA."

71. Yusoff, "Geologic Life," 784.

72. Peters, *Marvelous Clouds*, 2. I thank the anonymous reviewer for making this connection.

73. Zalasiewicz, *Earth After Us*, 163.

Bibliography

Adams, Frank D. "The Generation of Stones." In *The Birth and Development of the Geological Sciences*, 77–136. New York: Dover, 1938.
Adorno, Theodor W. *Aesthetic Theory*. Translated by R. Hullot-Kentor. Minneapolis: University of Minnesota Press, 1997.
———. "Über epische Naivetät." In *Noten zur Literatur*, 34–40. Frankfurt am Main: Suhrkamp, 1974.
Allen, Valerie. "Mineral Virtue." In *Animal, Vegetable, Mineral: Ethics and Objects*, edited by Jeffrey Jerome Cohen, 123–152. Washington, DC: Oliphaunt, 2012.
Alley, Richard B. *The Two-Mile Time Machine: Ice Cores, Abrupt Climate Change, and Our Future* Princeton, NJ: Princeton University Press, 2000.
Attanucci, Timothy J. "Stories from Earth: Adalbert Stifter and the Poetics of Earth History." PhD diss., Princeton University, Department of German, 2012.
Bachelard, Gaston. *The Formation of the Scientific Mind*. Translated by Mary MacAllester Jones. Manchester: Clinamen Press, 2002.
Bann, Stephen. "The Inscription in the Garden: Ian Hamilton Finlay and the Epigraphic Convention." *Apollo* 134, no. 354 (1991): 119–120.
Batten, Guinn. *The Orphaned Imagination: Melancholy and Commodity Culture in English Romanticism*. Durham, NC: Duke University Press, 1998.
Benjamin, Walter. *The Arcades Project*. Translated by Howard Eiland and Kevin McLaughlin. Cambridge, MA: Harvard University Press, 1999.
———. *Briefe*. 2 vols. Edited by Gershom Scholem and Theodor W. Adorno. Frankfurt am Main: Suhrkamp, 1966.
———. *The Correspondence of Walter Benjamin, 1910–1940*. Edited by Gershom Scholem and Theodor W. Adorno. Translated by Manfred R. Jacobson and Evelyn Jacobson. Chicago: University of Chicago Press, 1994.
———. *Gesammelte Schriften*. 7 vols. Edited by Rolf Tiedemann and Hermann Schweppenhäuser. Frankfurt am Main: Suhrkamp, 1972–1989.

———. *On Hashish*. Cambridge, MA: Harvard University Press, 2006.
———. *Selected Writings*. 4 vols. Edited by Marcus Bullock, Howard Eiland, and Michael Jennings. Cambridge, MA: Harvard University Press, 1996–2003.
Bennett, Jane. *Vibrant Matter: A Political Ecology of Things*. Durham, NC: Duke University Press, 2011.
Bez, Martin. *Goethes "Wilhelm Meisters Wanderjahre": Aggregat, Archiv, Archivroman*. Berlin: Walter de Gruyter, 2013.
Bisanz, Rudolf M. "The Birth of a Myth: Tischbein's "Goethe in the Roman Campagna." *Monatshefte* 80, no. 2 (Summer 1988): 187–199.
Blackall, Eric. *Adalbert Stifter: A Critical Study*. Cambridge: Cambridge University Press, 1948.
Blanchot, Maurice. "The Two Versions of the Imaginary." In *The Space of Literature*, translated by Ann Smock, 254–263. Lincoln: University of Nebraska Press, 1982.
Boehme, Hartmut. "Das Steinerne: Anmerkungen zur Theorie des Erhabenen aus dem Blick des 'Menschenfremdesten.'" In *Das Erhabene: Zwischen Grenzerfahrung und Größenwahn*, edited by Christine Pries, 119–141. Weinheim: VCH, Acta humaniora, 1989.
Boes, Tobias. "Reading the Book of the World: Epic Representation in the Age of Our Geophysical Agency." *Novel* 49, no. 1 (2016): 95–114.
Boes, Tobias, and Kate Marshall. "Introduction: Writing the Anthropocene." *Minnesota Review*, no. 83 (2014): 60–72.
Bonneuil, Christophe, and Jean-Baptiste Fressoz. *The Shock of the Anthropocene: The Earth, History and Us*. Translated by David Fernbach. London: Verso, 2015.
Boyle, Nicholas. *Goethe: The Poet and the Age. Vol. 2, Revolution and Renunciation (1790–1803)*. Oxford: Clarendon Press, 1999.
Brandt, Lindsey K. "*Tangled Up in Truths: German Literary Conceptions of Nature Between Romantic Science and Objective Empiricism*." PhD diss., University of North Carolina at Chapel Hill, Department of Germanic and Slavic Languages and Literatures, 2015.
Braungart, Georg. "Der Hauslehrer, Landschaftsmaler und Schriftsteller Adalbert Stifter besucht den Gletscherforscher Friedrich Simony. Hallstatt, im Sommer 1845." In *Bespiegelungskunst: Begegnungen auf den Seitenwegen der Literaturgeschichte*, edited by Georg Braungart et al., 101–118. Tübingen: Attempto, 2004.
———. "The Poetics of Nature: Literature and Constructive Imagination in the History of Geology." In *Inventions of the Imagination: Romanticism and Beyond*, edited by Richard T. Gray, Nicholas Halmi, Gary J. Handwerk, Michael A. Rosenthal, and Klaus A. Vieweg, 26–35. Washington: University of Washington Press, 2011.
Broch, Hermann. "James Joyce und die Gegenwart." In *Dichten und Erkennen*, 183–210. Zürich: Rhein Verlag, 1955.
Buckland, Adelene. *Novel Science: Fiction and the Invention of Nineteenth-Century Geology*. Chicago: University of Chicago Press, 2014.

Buell, Lawrence. *The Environmental Imagination; Thoreau, Nature Writing, and the Formation of American Culture*. Cambridge, MA: Harvard University Press, 1995.
Caillois, Roger. *Œuvres*. Paris: Gallimard, 2008.
Cameron, Dorothy. "Goethe—Discoverer of the Ice Age." *Journal of Glaciology* 5, no. 41 (1965): 751–754.
Campe, Rüdiger. "Form and Life in the Theory of the Novel." *Constellations* 18, no. 1 (2011): 53–66.
——. "The Rauschen of the Waves: On the Margins of Literature." *SubStance* 61 (1990): 21–38.
Caruth, Cathy. *Literature in the Ashes of History*. Baltimore, MD: Johns Hopkins University Press, 2013.
Caverero, Adriana. *Inclinations: A Critique of Rectitude*. Translated by Adam Sitze. Stanford, CA: Stanford University Press, 2016.
Celan, Paul. *Der Meridian: Endfassung, Vorstufen, Materialien*. Edited by Bernhard Böschenstein and Heino Schmull. Frankfurt am Main: Suhrkamp, 1999.
——. *Gesammelte Werke*. 7 vols. Edited by Beda Allemann and Stefan Reichert. Frankfurt am Main: Suhrkamp, 2000.
Chakrabarty, Dipesh. "The Climate of History: Four Theses." *Critical Inquiry* 35 (Winter 2009): 197–222.
Chambers, Ross. *Loiterature*. Lincoln: University of Nebraska Press, 1999.
Charpentier, Jean de. *Essai sur les glaciers*. Lausanne: Duclou, 1841.
Clark, Nigel. *Inhuman Nature: Sociable Life on a Dynamic Planet*. London: Sage, 2011.
Clark, Nigel, and Myra Hird. "Deep Shit." *O-Zone: A Journal of Object-Oriented* 1, no. 1 (2014): 44–52.
Clark, Timothy. "Blanchot and the End of Nature." *parallax* 16, no. 2 (2011): 20–30.
——. *Ecocriticism on the Edge: The Anthropocene as a Threshold Concept*. London: Bloomsbury, 2015.
——. "Scale: Derangement of Scale." In *Telemorphosis: Theory in the Era of Climate Change*, 1:149–167. Ann Arbor, MI: Open Humanities Press, 2011.
Coen, Deborah. *Climate in Motion: Science, Empire, and the Problem of Scale*. Chicago: University of Chicago Press, 2018.
Cohen, Jeffrey Jerome. *Stone: An Ecology of the Inhuman*. Minneapolis: University of Minnesota Press, 2015.
Cohen, Tom. "Anecographics: Climate Change and 'Late' Deconstruction." In *Impasses of the Post-Global: Theory in the Era of Climate Change*, vol. 2, edited by Henry Sussman, 32–57. Ann Arbor, MI: Open Humanities Press, 2012.
——. "'Climate Change' and the Rupture of Cultural Critique." *Comparative Literature: East and West* 15, no. 1 (2011): 13–48.
Cohen, Tom, and Claire Colebrook. Preface to *Twilight of the Anthropocene Idols*. Ann Arbor, MI: Open Humanities Press, 2016.
Crutzen, Paul J. "Geology of Mankind." *Nature* 415, no. 6867 (2002): 23.
Davies, Jeremy. *The Birth of the Anthropocene*. Oakland, CA: University of California Press, 2016.

Davis, Heather, and Zoe Todd. "On the Importance of a Date, or Decolonizing the Anthropocene." *ACME: An International E-Journal for Critical Geographies* 16, no. 4 (2017): 761–780.

Dawson, Graham. *Soldier Heroes: British Adventure, Empire and the Imagining of Masculinities*. London: Routledge, 1994.

Deleuze, Gilles. *Pure Immanence: Essays on a Life*. Translated by Anne Boyman. New York: Zone Books, 2001.

Deleuze, Gilles, and Felix Guattari. "Geophilosophy." In *What Is Philosophy?* Translated by Hugh Tomlinson and Graham Burchell, 85–116. New York: Columbia University Press, 1994.

DeMair, Jillian. "Geological Uncertainty and Poetic Creativity: The Material Agency of Findlinge for Droste-Hülshoff and Goethe." *Otago German Studies* 28 (2018): 41–72.

Derrida, Jacques. *The Beast and the Sovereign. Vol. 2*. Edited by Michel Lisse, Marie-Louise Mallet, and Ginette Michaud. Translated by Geoffrey Bennington. Chicago: University of Chicago Press, 2011.

———. *Of Grammatology*. Baltimore, MD: Johns Hopkins University Press, 1976.

Dimock, Wai Chee. "Literature for the Planet." *PMLA* 116, no. 1 (2001): 173–188.

Downing, Eric. *Double Exposures: Repetition and Realism in Nineteenth-Century German Fiction*. Stanford, CA: Stanford University Press, 2000.

Droste-Hülshoff, Annette von. "Die Mergelgrube." In *Historisch-kritische Ausgabe*. Vol. 1, edited by Winfried Woesler and Winfried Theiss, 50–53. Tübingen: Max Niemeyer, 1985.

Eiland, Howard, and Michael W. Jennings, *Walter Benjamin: A Critical Life*. Cambridge, MA: Harvard University Press, 2014.

Ellsworth, Elizabeth, and Jamie Kruse. "Introduction: Evidence: Making a Geologic Turn in Cultural Awareness." In *Making the Geologic Now: Responses to Material Conditions of Contemporary Life*, edited by Elizabeth Ellsworth and Jamie Kruse, 6–26. New York: Punctum Books, 2013.

Engelhardt, Wolf von. "Did Goethe Discover the Ice Age?" *Eclogae Geologicae Helvetiae* 92 (1999): 123–128.

———. *Goethes Weltansichten: Auch eine Biographie*. Weimar: Böhlau, 2007.

Fiorato, Pierfrancesco. "Zum Paradigma des Kommentars im Denken Walter Benjamins." *Jewish Studies Quarterly* 13, no. 3 (2006): 265–277.

Fogle, Douglas, ed. *Life on Mars: 55th Carnegie International*. Pittsburgh, PA: Carnegie Museum Store, 2008.

Ford, Thomas. "Romanthropocene." *Literature Compass* 15, no. 5, 2018. https://doi.org/10.1111/lic3.12464.

Frank, Manfred. "Steinherz und Geldseele. Ein Symbol im Kontext." In *Das kalte Herz*, 253–387. Frankfurt am Main: Suhrkamp, 1978.

Frederick, Samuel. *Narratives Unsettled: Digression in Robert Walser, Thomas Bernhard, and Adalbert Stifter*. Evanston, IL: Northwestern University Press, 2012.

Freud, Sigmund. "A Difficulty in the Path of Psychoanalysis." In *The Standard Edition of the Complete Psychological Works of Sigmund Freud*, vol. 17, translated by James Strachey, with Anna Freud, 137–144. London: Hogarth, 1975.
———. *Die Traumdeutung*. Vol. 2 of *Studienausgabe*. Frankfurt am Main: Fischer, 1982.
———. *The Interpretation of Dreams*. Translated by James Strachey. New York: Basic Books, 1955.
———. "One of the difficulties of psycho-analysis." Translated by Joan Riviere. In *Sigmund Freud: Collected Papers*, vol. 4, edited by Masud Khan, 347–356. London: Hogarth, 1971.
Fry, Paul. *Wordsworth and the Poetry of What We Are*. New Haven, CT: Yale University Press, 2014.
Gasperi, Carlos. "On the Language of Nature in Ludwig Tieck's *Der Runenberg*." *Monatshefte* 107, no. 3 (Fall 2015): 405–430.
Geulen, Eva. *Worthörig wider Willen: Darstellungsproblematik und Sprachreflexion in der Prosa Adalbert Stifters*. Munich: Ludicium, 1992.
Geulen, Hans. "Stifters Sonderlinge: Kalkstein und Turmalin." *Jahrbuch der Deutschen Schillergesellschaft* 17 (1973): 415–431.
Ghosh, Amitav. *The Great Derangement: Climate Change and the Unthinkable*. Chicago: University of Chicago Press, 2016.
———. "Petrofiction: The Oil Encounter and the Novel." *New Republic*. March 2, 1992, 29–34.
Gideon, Heidi. *Zur Darstellungsweise von Goethes Wilhelm Meisters Wanderjahre*. Göttingen: Vandenhoeck and Ruprecht, 1969.
Glissant, Édouard. *Philosophie de la Relation*. Paris: Gallimard, 2009.
Goethe, Johann Wolfgang von. *Collected Works*. 12 vols. Edited by Viktor Lange, Eric A. Blackall, Cyrus Hamlin, and Jane K. Brown. Princeton, NJ: Princeton University Press, 1994–1995.
———. *Goethes Gespräche: Eine Sammlung zeitgenössischer Berichte aus seinem Umgang auf Grund der Ausgabe und des Nachlasses*. 10 vols. Edited by Wolfgang Herwig. Zurich: Artemis Verlag, 1965–1987.
———. *Goethes Werke*. 143 vols. Weimar: Böhlau, 1887–1919.
———. *Sämtliche Werke, Briefe, Tagebücher und Gespräche*. 40 vols. Edited by Dieter Borchmeyer et al. *Frankfurt am Main*: Deutscher Klassiker Verlag, 1985–1999.
———. *Die Schriften zur Naturwissenschaft*. 29 vols. Edited by Dorothea Kuhn et al. Weimar: H. Böhlaus Nachfolger, 1947–2011.
Goodbody, Axel. "Melting Ice and the Paradoxes of Zeno: Didactic Impulses and Aesthetic Distanciation in German Climate Change Fiction." *Ecozon@: European Journal of Literature, Culture and Environment* 4, no. 1 (2013): 92–102.
Görner, Rüdiger. "Granit: Zur Poesie eines Gesteins." *Aurora* 53 (1993): 126–138.
Grimm, Jacob, and Wilhelm Grimm, *Deutsches Wörterbuch*. 33 vols. Munich: Deutscher Taschenbuch Verlag, 1984.
Grosz, Elizabeth. *Becoming Undone: Darwinian Reflections on Life, Politics, and Art*. Durham, NC: Duke University Press, 2011.

Haberkorn, Michaela. *Naturhistoriker und Zeitenseher: Geologie und Poesie um 1800; Der Kreis um Abraham Gottlob Werner (Goethe, A. v. Humboldt, Novalis, Steffens, G. H. Schubert)*. Frankfurt am Main: Peter Lang, 2004.

Handke, Peter. *Die Lehre der Sainte-Victoire*. Frankfurt am Main: Suhrkamp, 1980.

———. *Slow Homecoming*. Translated by Ralph Mannheim. New York: Farrar, Straus and Giroux, 1985.

Harris, Paul A., Richard Turner, and A. J. Noeck, eds. "Rock Record." Special issue. *SubStance* 47, no. 2 (2018).

Hartman, Geoffrey H. "Inscriptions and Romantic Nature Poetry." In *The Unremarkable Wordsworth*, edited by Donald G. Marshall, 31–46. Minneapolis: University of Minnesota Press, 1987.

Haubenreich, Jacob. "Poetry, Painting, Patchwork: Peter Handke's Intermedial Writing of *Die Lehre der Saint-Victoire*." *German Quarterly* 92, no. 2 (2019): 187–210.

Heise, Ursula K. *Sense of Place and Sense of Planet: The Environmental Imagination of the Global*. Oxford: Oxford University Press, 2008.

Henry Huxley, Thomas. "On a Piece of Chalk." In *The Major Prose of Thomas Henry Huxley*, edited by Alan P. Barr, 154–173. Athens, GA: University of Georgia Press, 1997.

Holmes, Tove. "An Archive of the Earth: Stifter's Geologos." *Seminar: A Journal of Germanic Studies* 54, no. 3 (2018): 281–307.

Hutton, Jane. "Erratic Imaginaries: Thinking Landscape as Evidence." In *Architecture in the Anthropocene: Encounters among Design, Deep Time, Science, and Philosophy*, edited by Etienne Turpin, 111–124. Ann Arbor: University of Michigan Press and Open Humanities Press, 2013.

Iovino, Serenella, and Serpil Oppermann. "Stories from the Thick of Things: Introducing Material Ecocriticism." Part I of "Theorizing Material Ecocriticism: A Diptych." *ISLE* 19, no. 3 (Summer 2012): 449–460.

Ireton, Sean. "Adalbert Stifter and the Gentle Anthropocene." In *Readings in the Anthropocene: The Environmental Humanities, German Studies, and Beyond*, edited by Sabine Wilke and Japhet Johnstone, 195–221. London: Bloomsbury, 2017.

———. "Between Dirty and Disruptive Nature: Adalbert Stifter in the Context of Nineteenth-Century American Environmental Literature." *Colloquia Germanica* 44, no. 2 (2011): 149–171.

Jackson, Myles W. "Natural and Artificial Budgets: Accounting for Goethe's Economy of Nature." *Science in Context* 7, no. 3 (1994): 409–431.

Jay Gould, Stephen. *Time's Arrow, Time's Cycle. Myth and Metaphor in the Discovery of Geological Time*. Cambridge, MA: Harvard University Press, 1987.

Joy, Eileen. "Notes Toward a Speculative Realist Literary Criticism." Online lecture. December 20, 2011. Svenska Twitteruniversitet. http://svtwuni.wordpress.com/2011/12/21/eileen.

Joyce, James. *Finnegans Wake*. London: Viking Press, 1939.

Ketelsen, Uwe. "Geschichtiches Bewußtsein als literarische Struktur: Zu Stifters Erzählung aus der Revolutionszeit 'Granit' (1848/52)." *Euphorion* 64 (1970): 306–325.

Koschorke, Albrecht. *Die Geschichte des Horizonts*. Frankfurt am Main: Suhrkamp, 1990.

Krüger, Tobias. *Discovering the Ice Ages: International Reception and Consequences for a Historical Understanding of Climate*. Translated by Ann M. Hentschel. Leiden: Brill, 2013.

Kuh, Emil. *Zwei Dichter Österreichs: Franz Grillparzer—Adalbert Stifter*. Pest: Hecknast, 1872.

Kuzniar, Alice. "Stones that Stare, or, the Gorgon's Gaze in Ludwig Tieck's *Der Runenberg*." In *Mimetic Desire. Essays on Narcissism in German Literature from Romanticism to Post Modernism*, edited by Jeffrey Adams and Eric Williams, 50–64. Columbia, SC: Camden House, 1995.

Lacan, Jacques. "Lituraterre." Translated by Dany Nobus. *Continental Philosophy Review* 46, no. 2 (2013): 327–334.

Latimer, Renate. "On Translating Stifter's 'Nachsommer.'" *Modern Austrian Literature* 12, no. 2 (1979): 67–79.

Latour, Bruno. *Facing Gaia: Eight Lectures on the New Climate Regime*. Translated by Catherine Porter. Cambridge, UK: Polity, 2017.

———. *Politics of Nature*. Cambridge, MA: Harvard University Press, 2009.

Lessing, Gotthold Ephraim. *Fables and Epigrams: With Essays on Fable and Epigram*. London: Hunt, 1825.

———. *Werke*. 25 vols. Edited by Julius Petersen and Waldemar von Olshausen. Berlin and Vienna: Deutsches Verlagshaus Bong, 1925–1935.

Leverett, Frederick Percival. *A New and Copious Lexicon of the Latin Language: Compiled Chiefly from the Magnum Totius Latinitatis of Facciolati and Forcellini and the German Works of Scheller and Luenemann*. Boston, MA: Wilkins and Carter and C. C. Little and James Brown, 1838.

Levine, Michael. *A Weak Messianic Power: Figures of a Time to Come in Benjamin, Derrida, and Celan*. New York: Fordham University Press, 2016.

Lovelock, James. *A Rough Ride to the Future*. London: Penguin UK, 2014.

Lukács, Georg. "Erzählen oder Beschreiben?" In *Probleme des Realismus 1: Essays über Realismus*, 197–242. Neuwied: Luchterhand, 1971.

Lyell, Charles. *Principles of Geology*. Vol. 1. Chicago: University of Chicago Press, 1990.

Lyon, John B. "Disorientation in Novalis or 'The Subterranean Homesick Blues.'" *Goethe Yearbook* 24 (2017): 85–103.

MacKay, John. *Inscription and Modernity: From Wordsworth to Mandelstam*. Bloomington: Indiana University Press, 2006.

Massey, Doreen. "Landscape as a Provocation: Reflections on Moving Mountains." *Journal of Material Culture* 11, nos. 1–2 (2006): 33–48.

McKibben, Bill. *Eaarth: Making a Life on a Tough New Planet*. New York: Times Books, 2010.

Menely, Tobias. "Anthropocene Air." *Minnesota Review*, 83 (2014): 93–101.

———. "'The Present Obfuscation': Cowper's Task and the Time of Climate Change." *PMLA* 127, no. 3 (May 2012): 477–492.

Menely, Tobias, and Jesse Oak Taylor. Introduction to *Anthropocene Reading: Literary History in Geologic Times*, edited by Tobias Menely and Jesse Oak Taylor, 9–36. Philadelphia: Penn State University Press, 2017.

Merola, Nicola. "Mediating Planetary Attachments and Planetary Melancholy: Lars von Trier's *Melancholia*." In *Design, Mediation, and the Posthuman*, edited by Dennis Weiss, Amy Propen, and Colbey Emmerson Reid, 249–267. London: Rowman and Littlefield, 2014.

Micheler, Werner. "Vulkanische Idyllen: Die Fortschreibung der Revolution mit den Mitteln der Naturwissenschaft bei Moritz Hartmann und Adalbert Stifter." In *Bewegung im Reich der Immobilität*, 472–495. Vienna: Böhlau, 2001.

Miezskowski, Jan. "The Writing is on the Wall." *Postmodern Culture* 21, no. 1 (2012). http://www.pomoculture.org/2013/09/03/the-writing-is-on-the-wall/.

Mirzoeff, Nicholas. "It's not the Anthropocene, It's the White Supremacy Scene, or, the Geological Color Line." In *After Extinction*, edited by Richard Grusin, 123–150. Minneapolis: University of Minnesota Press, 2018.

Mittelmeier, Martin. *Adorno in Neapel: Wie sich eine Sehnsuchtslandschaft in Philosophie verwandelt*. Munich: Siedler Verlag, 2013.

Moore, Jason. *Capitalism in the Web of Life*. London: Verso, 2017.

Moraru, Christian. *Reading for the Planet: Toward a Geomethodology*. Ann Arbor: University of Michigan Press, 2015.

Mortimer-Sandilands, Catriona, and Bruce Erickson, eds. *Queer Ecologies: Sex, Nature, Politics, Desire*. Bloomington: Indiana University Press, 2010.

Morton, Timothy. "Guest Column: Queer Ecology." *PMLA* 125, no. 2 (2010): 273–282.

———. *Hyperobjects: Philosophy and Ecology After the End of the World*. Minneapolis: University of Minnesota Press, 2013.

———. "Thinking Ecology: The Mesh, the Strange Stranger, and the Beautiful Soul." *Collapse* 6 (2010): 265–293.

Möseneder, Karl. "Stimmung und Erdleben. Adalbert Stifters Ikonologie der Landschaftsmalerei." *Adalbert Stifter: Dichter und Maler, Denkmalpfleger und Schulmann*, edited by Hartmut Laufhütte and Karl Möseneder, 18–57. Tübingen: Niemeyer, 1996.

Muenzer, Clark. "'Ihr ältesten, würdigsten Denkmäler der Zeit': Goethe's *Über den Granit* and His Aesthetics of Monuments." In *Ethik und Ästhetik: Werke und Werte in der Literatur vom 18. bis zum 20. Jahrhundert; Festschrift für Wolfgang Wittkowski zum 70. Geburtstag*, edited by Richard Fisher, 181–198. Frankfurt am Main: Peter Lang, 1995.

Müller-Sievers, Helmut. *The Science of Literature: Essays on an Incalculable Difference*. Berlin: De Gruyter, 2015.

Mumford, Lewis. *Technics and Civilization*. New York: Harper, 1934.

Neyrat, Frédéric. *The Unconstructable Earth*. New York: Fordham University Press, 2018.

Nixon, Rob. "The Anthropocene: Promise and Pitfalls of an Epochal Idea." *Edge Effects*. Updated October 12, 2019. http://edgeeffects.net/anthropocene-promise-and-pitfalls/.
———. *Slow Violence and the Environmentalism of the Poor*. Cambridge, MA: Harvard University Press, 2011.
Novalis. *Schriften*. Edited by Paul Kluckhohn and Richard Samuel. 4 vols. Stuttgart: Kohlhammer, 1960–1975.
Parikka, Jussi. *The Anthrobscene*. Minneapolis: University of Minnesota Press, 2014.
———. *A Geology of Media*. Minneapolis: University of Minnesota Press, 2015.
———. "The Geology of Media." *Atlantic*, October 11, 2013. https://www.theatlantic.com/technology/archive/2013/10/the-geology-of-media/280523/.
Peters, John Durham. *The Marvelous Clouds: Toward a Philosophy of Elemental Media*. Chicago: University of Chicago Press, 2015.
Phillips, Alexander. "Adalbert Stifter's Alternative Anthropocene: Reimagining Social Nature in Brigitta and Abdias." In *German Ecocriticism in the Anthropocene*, edited by Caroline Schaumann and Heather Sullivan, 65–85. New York: Palgrave Macmillan, 2017.
Piper, Andrew. "Mapping Vision: Goethe, Cartography, and the Novel." In *Spatial Turns: Space, Place, and Mobility in German Literary and Visual Culture*, edited by Jaimey Fisher and Barbara Mennel, 27–52. Amsterdam: Rodopi, 2010.
———. "Rethinking the Print Object: Goethe and the Book of Everything." *PMLA* 121, no. 1 (2006): 124–138.
Pörksen, Uwe. "Raumzeit: Goethes Zeitbegriff aufgrund seiner sprachlichen Darstellung geologischer Ideen und ihrer Visualisierung." In *Goethe und die Verzeitlichung der Natur*, edited by Peter Matussek, 101–127. Munich: Beck, 1998.
Povinelli, Elizabeth. *Geontologies: A Requiem to Late Liberalism*. Durham, NC: Duke University Press, 2016.
Powers, Elizabeth. "The Sublime, 'Über den Granit,' and the Prehistory of Goethe's Science." *Goethe Yearbook* 15 (2008): 35–56.
Prete, Ivano dal. "'Being the World Eternal.' The Age of the Earth in Renaissance Italy." *Isis* 105, no. 2 (2014): 292–317.
Proust, Marcel. *Sodom and Gomorrah*. Translated by C. K. Scott Moncrieff and Terence Kilmartin: New York: Modern Library, 2003.
Prutti, Brigitte. "Zwischen Ansteckung und Auslöschung: Zur Seuchenerzählung Bei Stifter—Die Pechbrenner Versus Granit." *Oxford German Studies* 37, no. 1 (2008): 49–73.
Ransom, John Crowe. "The Psychologist Looks at Poetry." *Virginia Quarterly Review* 11, no. 4 (1935): 575–592.
Reynard, Emmanuel. "Protecting Stones: Conservation of Erratic Blocks in Switzerland." In *Dimension Stone*, edited by Richard Prikryl, 3–8. London: Taylor and Francis, 2004.
Richter, Gerhard. *Thought-Images: Frankfurt School Writers' Reflections from Damaged Life*. Stanford, CA: Stanford University Press, 2007

———. "Toward a Politics of the Unusable." In *Walter Benjamin and The Corpus of Autobiography*, 231–246. Detroit: Wayne State University Press, 2000.

Rigby, Kate. "'Mines aren't really like that': German Romantic Undergrounds Revisited." In *German Ecocriticism in the Anthropocene*, edited by Caroline Schaumann and Heather Sullivan, 111–128. New York: Palgrave Macmillan, 2017.

———. *Topographies of the Sacred: The Poetics of Place in European Romanticism*. Charlottesville: University of Virginia Press, 2004.

Roof, Judith. "From Protista to DNA (and Back Again): Freud's Psychoanalysis of the Single-Celled Organism." In *Zoontologies: The Question of the Animal*, edited by Cary Wolfe, 1–20. Minneapolis: University of Minnesota Press, 2003.

Rosei, Peter. "Versuch über Stifter und einige Schriftsteller der Gegenwart." *Literatur und Kritik*, no. 103 (1976): 161–167.

Safranski, Rüdiger. *Romanticism: A German Affair*. Translated by Robert E. Goodwin. Evanston, IL: Northwestern University Press, 2014.

Santner, Eric L. *Stranded Objects: Mourning, Memory, and Film in Post-War Germany*. Ithaca, NY: Cornell University Press, 1993.

Schellenberger-Diederich, Erika. *Geopoetik: Studien zur Metaphorik des Gesteins in der Lyrik von Hölderlin bis Celan*. Bielefeld: Aisthesis, 2006.

Schestag, Thomas. "Interpolationen: Benjamins Philologie." *philo:xenia* 1 (2009): 33–99.

Sabine Schneider, "Kulturerosionen. Stifters prekäre geologische Übertragungen." In *Figuren der Übertragung: Adalbert Stifter und das Wissen seiner Zeit*, edited by Michael Gamper and Karl Wagner, 249–269. Zurich: Chronos, 2009.

Schnyder, Peter. "Die Dynamisierung des Statischen: Geologisches Wissen bei Goethe und Stifter." *Zeitschrift für Germanistik* 19, no. 3 (2009): 540–555.

———. "Schrift—Bild—Sammlung—Karte: Medien geologischen Wissens in Stifters *Nachsommer*." In *Figuren der Übertragung: Adalbert Stifter und das Wissen seiner Zeit*, edited by Michael Gamper and Karl Wagner, 235–248. Zurich: Chronos, 2009.

Schubert, Gotthilf Heinrich. *Ansichten von der Nachtseite der Naturwissenschaft*. Dresden: Arnoldische Buchhandlung, 1808; reprint Eschborn: Dietmar Klotz, 1992.

Schutjer, Karen. "Beyond the Wandering Jew: Anti-Semitism and Narrative Supersession in Goethe's 'Wilhelm Meisters Wanderjahre.'" *German Quarterly* 77, no. 4 (2004): 389–407.

Schwab, Gabriele. *Haunting Legacies: Violent Histories and Transgenerational Trauma*. New York: Columbia University Press, 2010.

Sebald, W. G. "Bis an den Rand der Natur: Versuch über Stifter." In *Die Beschreibung des Unglücks: Zur österreichischen Literatur von Stifter bis Handke*, 15–37. Salzburg: Residenz, 1985.

Selz, Jean. "An Experiment by Walter Benjamin." In Walter Benjamin, *On Hashish*, edited and translated by Howard Eiland, 147–155. Cambridge, MA: Harvard University Press, 2006.

Silverberg, David. "Reflections: The Death of Gallium." *Asimov's Science Fiction* 32, no. 6 (June 2008): 8–10.

Sng, Zachary. "Not Forgotten: On Stifter and Peirce." *MLN* 121, no. 3 (2006): 631–646.

Sorg, Klaus Dieter. *Gebrochene Teleologie: Studien zum Bildungsroman von Goethe bis Thomas Mann.* Heidelberg: Winter, 1983.

Staiger, Emil. *Goethe.* 3 vols. Zurich: Atlantis Verlag, 1959.

Steffen, Will, Paul Crutzen, and John R. McNeill. "The Anthropocene: Are Humans Now Overwhelming the Great Forces of Nature?" *AMBIO: A Journal of the Human Environment* 36, no. 4 (2007): 614–621.

Steffens, Heinrich. *Was ich erlebte Aus der Erinnerung niedergeschrieben von Heinrich Steffens.* Vol. 3. Breslau: Josef Mar, 1841.

Stifter, Adalbert. "Granite." Translated by Jeffrey L. Sammons. In *German Novellas of Realism, vol. 1,* edited by Jeffrey L. Sammons, 7–34. New York: Continuum, 1989.

———. *Indian Summer.* Translated by Wendell Frye. New York: Peter Lang, 1985.

———. *Werke.* 4 vols. Edited by Uwe Japp and Hans Joachim Piechotta. Frankfurt am Main: Insel, 1978.

———. *Werke und Briefe: Historisch-Kritische Gesamtausgabe.* 10 vols. Edited by Alfred Doppler and Wolfgang Frühwald. Stuttgart: Kohlhammer, 1978–2008.

Strowick, Elisabeth. "'Dumpfe Dauer': Langeweile und Atmosphärisches bei Fontane und Stifter." *Germanic Review: Literature, Culture, Theory* 90, no. 3 (2015): 187–203.

———. "Poetological-Technical Operations: Representation of Motion in Adalbert Stifter." *Configurations* 18, no. 3 (Fall 2010): 273–289.

Sullivan, Heather. "Collecting the Rocks of Time: Goethe, the Romantics and Early Geology." *European Romantic Review* 10, no. 3 (1999): 341–370.

———. "Dirty Nature: Ecocriticism and Tales of Extraction—Mining and Solar Power—in Goethe, Hoffmann, Verne, and Eschbach." *Colloquia Germanica* 44, no. 2 (2014): 133–148.

———. "Faust's Mountains: An Ecocritical Reading of Goethe's Tragedy and Science." In *Heights of Reflection: Mountains in the German Imagination from the Middle Ages to the Twenty-First Century,* edited by Sean Moore Ireton and Caroline Schaumann, 116–133. Rochester, NY: Camden House, 2012.

———. "Material Ecocriticism and the Petro-Text." In *The Routledge Companion to the Environmental Humanities,* edited by Ursula K. Heise et al., 414–423. London: Routledge, 2017.

———. "Organic and Inorganic Bodies in the Age of Goethe: An Ecocritical Reading of Ludwig Tieck's 'Rune Mountain' and the Earth Sciences." *ISLE* 10, no. 2 (2003): 21–46.

———. "Ruins and the Construction of Time: Geological and Literary Perspectives in the Age of Goethe." *Studies in Eighteenth-Century Culture* 30 (2001): 1–30.

Swales, Martin. *The German Novelle.* Princeton, NJ: Princeton University Press, 1977.

Swales, Martin, and Erika Swales. *Adalbert Stifter: A Critical Study.* Cambridge: Cambridge University Press, 1984.

Szerszynski, Bronislaw. "The End of the End of Nature: The Anthropocene and the Fate of the Human." *Oxford Literary Review* 34, no. 2 (2012): 165–184.

Thüsen, Joachim von der. "Goethes Vulkane: Naturgewalt in der *Italienischen Reise*." *Neophilologus* 87, no. 2 (2003): 265–280.

Tieck, Ludwig. "Der Runenberg." In *Phantasus*, edited by Manfred Frank, 365–392. Frankfurt am Main: Deutscher Klassiker, 1985.

Tobias, Rochelle. *The Discourse of Nature in the Poetry of Paul Celan: The Unnatural World*. Baltimore, MD: Johns Hopkins University Press, 2006.

Trexler, Adam. *Anthropocene Fictions: The Novel in a Time of Climate Change*. Charlottesville: University of Virginia Press, 2015.

Twellman, Marcus. "Bleibende Stelle. Zu Stifters 'Granit.'" *Zeitschrift für deutsche Philologie* 126, no. 4 (2007): 226–243.

Uglietti, Chiara et al. "Widespread Pollution of the South American Atmosphere Predates the Industrial Revolution by 240 y." *Proceedings of the National Academy of Sciences* 112, no. 8 (2015): 2349–2354.

Vischer, Friedrich Theodor. "Aesthetik oder Wissenschaft des Schönen." In *Romantheorie: Texte vom Barock bis zur Gegenwart*, edited by Hartmut Steinecke and Fritz Wahrenburg, 364–368. Stuttgart: Reclam, 1999.

Vizenor, Gerald. *Survivance. Narratives of Native Presence*. Edited by Gerald Vizenor. Lincoln: University of Nebraska Press, 2008.

von Engelhardt, Wolf. *Goethe im Gespräch mit der Erde: Landschaft, Gesteine, Mineralien und Erdgeschichte in seinem Leben und Werk*. Weimar: Hermann Böhlaus Nachfolger, 2003.

Wackenroder, Wilhelm Heinrich. *Sämtliche Werke und Briefe*. 2 vols. Edited by Silvio Vietta and Richard Littlejohns. Heidelberg: Winter Verlag, 1991.

Wagenknecht, Christian. "Marxistische-Epigrammatik: Zu Bertolt Brechts 'Kriegsfibel.'" *Emblem und Emblematikrezeption: Vergleichende Studien zur Wirkungsgeschichte* 16, no. 20 (1978): 543–559.

Wagner, Irmgard. "Der Findling: Erratic Signifiers in Kleist and Geology." *German Quarterly* 65, no. 3 (1991): 281–295.

Wagner, Lori. "Schick, Schichten, Geschichte: Geological Theory in Stifter's Bunte Steine." *JASILO* 2 (1995): 17–41.

Watchman, Renae, Carrie Smith, and Markus Stock. "Building Transdisciplinary Relationships: Indigenous and German Studies." *Seminar: A Journal of Germanic Studies* 55, no. 4 (2019): 309–327.

Waters, Colin N., et al. "The Anthropocene is functionally and stratigraphically distinct from the Holocene." *Science* 351, no. 6269 (2016): 137–147.

Webb, Jonathan. "Tedium, Tragedy and Tar: The Slowest Drops in Science." *BBC News*. July 26, 2014. https://www.bbc.com/news/science-environment-28402709.

Wenzel, Jennifer. "Stratigraphy and Empire: *Waiting for the Barbarians*, Reading Under Duress." In *Anthropocene Reading: Literary History in Geologic Times*, edited by Tobias Menely and Jesse Oak Taylor, 167–183. Pennsylvania: Penn State University Press, 2017.

Westphal, Bertrand. *Geocriticism: Real and Fictional Spaces*. Translated by Robert T. Tally. New York: Palgrave Macmillan, 2011.

Wiedorn, Michael. *Think Like an Archipelago: Paradox in the Work of Édouard Glissant*. Albany: State University of New York Press, 2017.

Wild, Michael. *Wiederholung und Variation im Werk Adalbert Stifters*. Würzburg: Könighausen und Neumann, 2001.

Wilke, Sabine. "Toward an Environmental Aesthetics: Depicting Nature in the Age of Goethe." In *Goethe's Ghosts: Reading and the Persistence of Literature*, edited by Simon Richter and Richard Block, 262–275. New York: Camden House, 2013.

Woodard, Ben. *On an Ungrounded Earth: Towards a New Geophilosophy*. Brooklyn, NY: Punctum, 2013.

Woods, David. "Cosmic Passions." In *Deep Times, Dark Times: On Being Geologically Human*, 36–46. New York: Fordham University Press, 2018.

Wyder, Margit. "Goethes geologische Passionen: Vom Alter der Erde." *Goethe Jahrbuch* 125 (2008): 136–146.

———. "Gotthard, Gletscher und Gelehrte: schweizer Anregungen zu Goethes naturwissenschaftlichen Studien." In *Goethe und die Schweiz*, edited by Oliver Ruf, 23–110. Hanover: Wehrhahn, 2013.

Yusoff, Kathryn. "Geologic Life: Prehistory, Climate, Futures in the Anthropocene." *Environment and Planning D: Society and Space* 31, no. 5 (2013): 779–795.

Zalasiewicz, Jan. *The Earth After Us: The Legacy That Humans Will Leave in the Rocks*. Oxford: Oxford University Press, 2008.

Ziolkowski, Theodore. *German Romanticism and Its Institutions*. Princeton, NJ: Princeton University Press, 1990.

Index

Adorno, Gretel, Benjamin's letter to, 102, 105, 110, 154n18
Adorno, Theodor, *Aesthetic Theory*, 9, 95–96
Aesthetic Geology (Heringman), 5, 38
Agassiz, Louis, 45
Anthropocene, 43–44, 95, 116; agency of, 20, 124; colonialism and, 3, 124, 143n10; dating of, 18–19, 118, 126; destratification, 14, 113, 119, 128–129, 136; literature and, 10–11; the novel and, 119–120; romanticism and, 19–21, 27, 34
"Auf der Mauer stand mit Kreide" (Brecht), 129–131
Arcades Project (Benjamin), 98–99

Bachelard, Gaston, *The Formation of the Scientific Mind*, 47, 149n42
Becoming Undone (Grosz), 23–24
Benjamin, Walter: Adorno, Gretel, letter to, 102, 105, 110, 154n18; *Arcades Project*, 98–99; *Berlin Childhood*, 93, 103–104, 106, 111; "Commentary on Poems by Brecht," 97–99, 125–137; *The Concept of Art Criticism*, 31–32; "Downhill," 101–102, 107–113; *Erschütterung*, 93–99, 108–110; "Hashish in Marseilles," 100–104; Ibiza, 97, 102–110; on lapidary style, 14, 118, 128–133; "Naples," 105–106; *One-Way Street*, 105, 133, 154n18; "On the Concept of History," 113; porosity, 105–106, 114; Proust, 107, 110; shock, 12, 14, 71–72, 89, 93–99, 107, 112, "Stifter II," 96–97; "Task of the Translator," 98, 133; thought image (*Denkbild*), 104–106; "To the Planetarium," 112; survival, 132–133; walking, 100–102, 111–112

Bennett, Jane, *Vibrant Matter*, 20, 26–27, 44, 70–72
Berlin Childhood (Benjamin), 93, 103–104, 106, 111
blocks: epistemological, 43–47, 55, 59, 149n42; granite, 2–4, 13, 38–46, 48–55; stumbling, 44, 55
Bonneuil, Christoph, *The Shock of the Anthropocene*, 7, 95
Bradford, Mark, *Help Us*, 115–116, 122
Braungart, Georg, 5
Brecht, Bertolt: "Auf der Mauer stand mit Kreide," 129–131, 136; "Deutsche Kriegsfibel," 129–132; "Gegen Verführung," 97–98; "Hoch Lenin," 134; *War Primer* (*Kriegsfibel*), 130
Broch, Hermann, 55
Buckland, Adelene, 5
Buffon, Comte de (Georges-Louis Leclerc), 36

Caillois, Roger, 25
Campe, Rüdiger, 28, 120
Carnegie Museum of Natural History, 115–116
Caruth, Cathy, *Literature in the Ashes of History*, 7
Cavarero, Adriana, *Inclinations*, 4, 39
Celan, Paul: enrichment of the German language, 9–10; *Meridian* speech, 4–5
Chakrabarty, Dipesh, 66
chalk, 12, 135–136
Clark, Nigel, *Inhuman Nature*, 20, 27, 43–44
Clark, Timothy, 44, 94; *Ecocriticism on the Edge*, 126–127, 132
Cohen, Jeffrey Jerome, *Stone*, 24, 117

"Commentary on Poems by Brecht" (Benjamin), 97–99
The Concept of Art Criticism (Benjamin), 31–32
colonialism, 3, 19, 124
Cretaceous Period, 135
Crutzen, Paul, 19

Davies, Jeremy, *Birth of the Anthropocene*, 120–121, 124–125, 128–129
Deleuze, Gilles, "Pure Immanence: A Life," 27
derangement, 14, 21, 93–94, 125–127
Derrida, Jacques, 119, 132
"Deutsche Kriegsfibel" (Brecht), 129–132
Dimock, Wai Chee, 133
discards, 13, 62, 72, 89, 90
"Downhill" (Benjamin), 101–102, 107–113
Downing, Eric, 77
Droste-Hülshoff, Annette, "The Marl Pit," 40

Eaarth (McKibben), 1, 10
the Earth: unarticulated, 100, 107, 113–114, 153n13; unregulated, 14, 100, 114, 153n13; unreliable, 68, 94, 120; the whole earth, 29, 69
Engelhardt, Wolf von, 58, 62
epigram, 116–122, 124–125, 130–133
"erratic," the, 6–7, 43, 54, 65
Erschütterung (shock), 12, 14, 70–72, 83, 89, 93–99, 108–109. *See also* shock
extinction, 135–137

Faust II (Goethe), 58, 149n66
Finnegan's Wake (Joyce), 134
The Formation of the Scientific Mind (Bachelard), 47, 149n42
fossil capitalism, 15, 124, 137
Freiberg Mining Academy, 2, 25, 46, 53
Fressoz, Jean-Baptiste, *The Shock of the Anthropocene*, 7, 95
Freud, Sigmund, 47: *Beyond the Pleasure Principle*, 136; "A Difficulty in the Path of Psychoanalysis," 8–9; *Interpretation of Dreams*, 84
Fry, Paul, 123, 155n35

"Gegen Verführung" (Brecht), 97–98
geoaffect, 26, 94
geologic turn, 37
geological unconscious, 7–8
"Geologische Probleme" (Goethe), 64
geology, 5–6, 88; geological and geologic, 6; geology of media, 13
A Geology of Media (Parikka), 18, 20, 22, 32
geomethodology, 22
Geontologies (Povinelli), 26
geotrauma, 8, 29
"Gespräch über die Bewegung von Granitblöcken durch Gletscher" (Goethe), 62–66
Ghosh, Amitav, 17, 125
Glissant, Édouard, 102
Goethe, Johann Wolfgang von: discovery of ice age, 44, 53–54, 61; Engelhardt, Wolf von on, 62; epistemological blocks, 45–48; erratic terminology, 44–48; *Faust II*, 58, 149n66; geological collection, 40–41; "Geologische Probleme," 64; "Gespräch über die Bewegung von Granitblöcken durch Gletscher," 62–66; "Granitarbeiten in Berlin" 3–4, 48–53; granite basin (*Granitschale*), 3, 49–50, 54; Karlsbad, 58; Markgrafenstein, 48–51, 53, 60, 63; "On Granite," 37–39, 59; *Roman über das Weltall* (*Cosmic novel*), 36; Tischbein's portrait, 39; "Toward a Theory of Weather," 58; "Umherliegender Granit," 64; *Wilhem Meister's Journeyman Years*, 41–43, 54–66
Gould, Stephen Jay, 8, 144n14
"Granitarbeiten in Berlin" (Goethe), 3–4, 48–53
Granite (Stifter): absence of granite in, 85; blocks in, 84–85; diminution, 86; erosion, 82, 85–86; plague, 78, 80–84, 86, 89, 91; revision of *The Pitch Burners*, 78, 80–84, 87–88; spectral stones, 82–83
Grosz, Elizabeth, *Becoming Undone*, 23–24

Handke, Peter, *The Lesson of Mont Sainte-Victoire*, 79
Hartman, Geoffrey, "Inscription and Romantic Nature Poetry," 116, 121–123
"Hashish in Marseilles" (Benjamin), 100–104
Heise, Ursula, 11
Heringman, Noah, *Aesthetic Geology*, 5, 38
"Hoch Lenin" (Brecht), 134
Humboldt, Alexander von, 47–48, 86
humiliation, 8–13
Huxley, Thomas Henry, "On a Piece of Chalk," 135–136
Hyperobjects (Morton), 8–9, 43

inclinations, toward the inorganic, 3–6, 8, 36–40, 55, 72, 117–118
Indian Summer (Stifter), 68–71, 73–77, 86–87, 120–121; derogatory connotations of title, 150n5
Inscription and Modernity (MacKay), 121–124, 135

INDEX

"Inscription and Romantic Nature Poetry" (Hartman), 116, 121–123

Joyce, James, *Finnegan's Wake*, 134

Kuzniar, Alice, 26, 33

Lacan, Jacques, 26; "Lituraterre," 11, 91
lapidary, 14, 117–118, 128–133
Latour, Bruno, 5–6, 57, 94, 116
Lessing, Gotthold Ephraim, 130–131
The Lesson of Mont Sainte-Victoire (Handke), 79
Levine, Michael, 113
Lewis, Mumford, 21
Limestone (Stifter), 78–79
lithography, 32
"Lituraterre" (Lacan), 11, 91
Lukács, Georg, 70
Lyell, Charles, *Principles of Geology*, 69, 118

MacKay, John, *Inscription and Modernity*, 121–124, 135
Many-Colored Stones (Stifter), 69–72, 77–79, 81, 85, 89, 99, 151n30, 152n42
"The Marl-Pit," (Droste-Hülshoff), 40
Massey, Doreen, 43
McKibben, Bill, *Eaarth*, 1–2, 10
Menely, Tobias, 113, 118
Meridian speech (Celan), 4–5
mineral community, 123, 135–136
mineral gaze, 31–34
minerality, 4, 16, 123
mining, 2, 10, 21–25, 31–34
Moraru, Christian, 11–12
Morton, Timothy, *Hyperobjects*, 8–9, 21, 43, 128
Movement (II) (Stifter), 74
Müller-Sievers, Helmut, 42

Nancy, Jean-Luc, 17
"Naples" (Benjamin), 105–106
Newcomen engine, 19–20, 22
Nicolson, Marjorie, 6
Nixon, Rob, 34; *Slow Violence*, 65, 94
Novalis (Friedrich von Hardenberg), 19–20, 31–32

"On a Piece of Chalk" (Huxley, Thomas Henry), 135–136
"On Granit" (Goethe), 37–39
"On the Concept of History" (Benjamin), 113
One-Way Street (Benjamin), 105, 133, 154n18

Parikka, Jussi, *A Geology of Media*, 18, 20, 22, 32

Peters, John Durham, 136
petrofiction, 13, 17–18
Piper, Andrew, 5, 38, 62
The Pitch Burners (Stifter), 67–68, 90; revision into *Granite*, 78, 80–84, 87–88
planetary dysphoria, 30
Povinelli, Elizabeth, *Geontologies*, 26
Principles of Geology (Lyell), 69, 118

queer ecology, 37, 147n4

"Resolution and Independence" (Wordsworth), 38
Rigby, Kate, 21
Roman über das Weltall (*Cosmic novel*) (Goethe), 36
Rune Mountain (Tieck): dark ecology of, 21; geotrauma, 29–30; mineral gaze, 26, 31–34; noise, 23, 26, 28–31

Santner, Eric, *Stranded Objects*, 78
Saussure, Horace Bénédict de, 46
Schubert, Gotthilf Heinrich von, 25
Schwab, Gabriele, 7
Sebald, W. G., 70–71, 79
shock, 14, 70–73, 89, 93–99, 107, 112, 136. See also *Erschütterung*
The Shock of the Anthropocene (Bonneuil and Fressoz), 7, 95
Simony, Friedrich, 67, 69
Slow Violence (Nixon), 65, 94
Staiger, Emil, 55
Steffens, Heinrich, 2, 25, 146
"Stifter II" (Benjamin), 96–97
Stifter: Benjamin, Walter on, 71–72; deep time, 68–70, 76; earth-magnitude observation, 69–70, 76; erosion, 72, 74, 92; *Granite*, 77–92; Handke, Peter on, 79; *Indian Summer*, 68–71, 73–77, 86–87, 120–121; *Limestone*, 78–79; Lukács, Georg on, 70; *Many-Colored Stones*, 69–72, 77–79, 81, 85, 89, 99, 151n30, 152n42; *Movement (II)*, 74; *The Pitchburners*, 67–68, 78, 80–84, 87, 90–91; response to 1848 uprisings, 78, 80, 88–89; revisionary practices, 72, 89–90; Simony, Friedrich and, 67, 69; textile metaphors, 78–82; Vienna, 86–87; "A Walk through the Catacombs," 86–87
Stranded Objects (Santner), 78
stratigraphy, 5–6, 18, 38, 75–76; the Anthropocene and, 119–125; Benjamin, 129; Brecht, 134–137; destratification, 14, 113, 119, 128–129, 136; Goethe, 37–38; Handke, Peter, 79; Stifter, 89–90
Sullivan, Heather, 9, 17, 21, 25–26, 30, 58

Swales, Erika, 82
Swales, Martin, 70–71, 82

Taylor, Jesse Oak, 118
Tieck, Ludwig: *Rune Mountain*, 12–13, 15, 24–35; visit to Kemlas mine with Wackenroder, 12, 18–19, 22–24
tuff, 97

unconformities, 3, 55, 90

Vibrant Matter (Bennett), 20, 26–27, 44, 70–72
Vischer, Friedrich Theodor, 80
Vizenor, Gerald, 132–133
Vogl, Joseph, 5

Wackenroder, Wilhelm Heinrich, 18, 22–23
"A Walk through the Catacombs" (Stifter), 86–87

War Primer (Brecht), 130
Weichsel Ice Age, 44
Wenzel, Jennifer, 129
Werner, Abraham Gottlob, 2, 37, 45–46, 60
Wilhem Meister's Journeyman Years (Goethe): anti-Semitism of, 42, 148n26; drama of things, 57–58; geological problems in, 59–62; renunciation in, 59–60; wandering in, 55–56, 65
Woodard, Ben, 99
Wordsworth, William, "Resolution and Independence," 38

Yusoff, Kathryn, 18, 136

Zalasiewicz, Jan, 119, 121, 123, 137
Zantop, Susanne, 2–3
Zentralanstalt für Meteorologie und Geodynamik (ZAMG), 69
Ziolkowski, Theodor, 21–23, 31

Jason Groves is Assistant Professor of Germanics at the University of Washington. He is co-translator of Werner Hamacher's *Minima Philologica*.